I0065150

Thank you to Ted Ingraham and Elton Hall, who reviewed the manuscripts and provided editorial assistance in producing this publication.

Front cover: A. J. Fullam's American Stencil Tool Works. Paul Kebabian discusses the print on page 12.

Back cover: A Dewey & Newton, twenty-two inch, beech fore plane with a Moulson Brothers double iron. The story of this Dewey & Newton plane begins on page 20.

Photograph ●f Paul Kebabian courtesy of Ted Ingraham
Production: Patty MacLeish, Ideas into Print, Newport, R.I.

The articles in this publication had previously been published in The Chronicle, the ACTIVE Newsletter, *and* Scrapbook *and are reprinted here with permission.*

Paul Kebabian
Writings on Tools & Toolmakers

Paul Kebabian: Writings on Tools & Toolmakers *was produced on the occasion of the 2003 Early American Industries Association Annual Meeting in Burlington, Vermont, through a generous grant from Pete and Donna Hathaway.*

Paul Kebabian

Foreword

For serious tool collectors one of the most important parts of their tool collections is the written material that they use to research and learn about the artifacts they collect. In a large sense, every collector who regularly consults the library is continuing the work of the original author by adding additional observations and perhaps differing points of view. Recording and communicating one's observations to others through the written word can be the most important legacy any collector can hope to attain for their many years of study. The reprinted articles contained in this volume are only a portion of the numerous contributions to the field of early American tool studies made by Paul Kebabian.

Paul began his tool collecting adventure in the late 1960s and has been able to assemble an enormous collection of American-made hand tools reaching from the beginning of the eighteenth to the dawn of the twentieth century. His travels in search of tools have led him throughout New England, and the tags attached to each and every tool that he has collected read like a who's who of New England antique shops. While Paul was fortunate enough to get in on the "good old days" of collecting, when planes could still be purchased at auction for a couple of dollars for a bushel basket full, his diverse interest and keen eye have kept him looking for that "special tool." His collection ranges from the rare and unusual to the commonplace, reflecting his quest for a full understanding of the field.

By vocation Paul is a librarian, a fact that he never fails to take advantage of. He worked at the New York Public Library, University of Baghdad, and the University of Florida, and finished his career as a librarian as the Director of Library Services at the University of Vermont. Over the years, he has spent countless hours chasing down patent records and locating supporting historical information. As a founding member of ACTIVE (Antique Crafts and Tools in Vermont) in the early 1970s, Paul served as the editor of both the newsletter and *Scrapbook*, using much of the material he had researched to fill the pages of the group's publications. His ability as a wordsmith has allowed him to communicate his discoveries to other collectors in concise, clearly written articles that are a pleasure to read. The author of several important books on American tools, Paul is a longtime Early American Industries Association member—and its president from 1973-1976—and a contributor to *The Chronicle*.

Throughout his career collecting tools, researching them, and then writing about them with clarity and insight, Paul has clearly demonstrated how to share important collections and the collector's personal observations about them. While some of his articles that appeared in the ACTIVE newsletter and *Scrapbook* were later reprinted in *The Chronicle*, most of them were not and are unknown to many of us. It seemed only fitting that this year, when the EAIA is holding its annual meeting in Vermont, that we honor Paul by publishing this selection of his writings to share with a wider audience.

Ted Ingraham, May 2003

Paul Kebabian
Writings on Tools & Toolmakers

Contents

Some Eagle Square Special Orders, 1870–1875 & the Bridge Builder's Square

A voluminous record of the Eagle Square Company, covering over seventy years of the firm's business activity (1859–1874 as the Eagle Square Co.; 1874–1916 as the Eagle Square Manufacturing Co.; and 1916–1932 as a unit of the Stanley Rule and Level Co. of New Britain, Connecticut) forms a part of the manuscript collections in the Special Collections Department of the University of Vermont's Bailey/Howe Library.

Included in the material are inventories, cash books, journals, books of check stubs, bank books, ledgers, payroll books, forty-nine cartons containing letters and copies of correspondence, and files of purchase orders for accounts payable and receivable. The correspondence files cover chiefly the period 1903–1932. Twenty-seven "letter books" of outgoing correspondence (1864–1903) and five "books" of originals of incoming letters (1869–1878) form another part of the records.

The many copy books of outgoing correspondence are unfortunately largely illegible. They consist of bound volumes of a light, translucent paper with the impressed copies of letters. An outgoing letter, hand written in ink, was inserted under one of the dampened, blank tissue pages and the book put in a copy press (similar to, and often mistaken for a bookbinder's press and of the type of copy press planemaker Leonard Bailey was manufacturing in this era). The pressure of the copy press transferred some of the ink from the original letter to the tissue, normally sufficient to provide a legible duplicate. But moisture, blurring, and deterioration of the inked impressions over the past eighty to one hundred twenty years has made most of these letter copies unreadable.

The letter books of incoming correspondence consist of blank "books" with paper stubs at the spine on which the original letters were tipped in with glue. The information which follows is for the most part based on selected information from letters in two of the books covering 1870–1875. These deal principally with incoming special orders—rather than the large consignment orders which made up the bulk of the firm's production of squares.

Some general comments are worth noting: the handwriting of most letters is very legible; almost all the letters are on what are now described as half-sheets (approximately four inches by eight inches)—brevity in letter writing was the watchword, and Mr. Mattison of Eagle Square several times made pencilled notes concerning responses to the orders, where he had to ask for more complete specifications; and mail service was rapid—a letter, reply, and follow-up letter between South Shaftsbury and Albany, New York, for example, often took but three successive days because train service was so frequent.

Steel and Iron Squares.

NOS.	NAMES.	WIDTH.	PRICE.	DESCRIPTION.
100	Cast Steel, Improved,	2 in.	$66.00	$\frac{1}{8}$, $\frac{1}{12}$, $\frac{1}{16}$, $\frac{1}{2}$, $\frac{1}{4}$, with brace, 8 square and $\frac{1}{100}$th scale, and Essex's new board measure, giving feet and inches in full.
1	Cast Steel for Drafting,	2 "	48.00	$\frac{1}{8}$, $\frac{1}{12}$, $\frac{1}{8}$, $\frac{1}{4}$, board and brace measure, 8 square, and $\frac{1}{100}$th scale.
2	Cast Steel Finish,	2 "	44.00	$\frac{1}{8}$, $\frac{1}{12}$, $\frac{1}{8}$, $\frac{1}{4}$, board and brace measure, and 8 square scale.
2½	Framing,	2 "	40.00	$\frac{1}{8}$, $\frac{1}{4}$, both sides and edges.
3	Sup. Sup. Extra,	2 "	35.00	$\frac{1}{8}$, $\frac{1}{12}$, $\frac{1}{8}$, $\frac{1}{4}$, board and brace measure.
4	Super Extra,	2 "	33.50	$\frac{1}{12}$, $\frac{1}{8}$, $\frac{1}{4}$, board and brace measure, extra figures on inner edge of body.
5	Extra,	2 "	32.50	$\frac{1}{12}$, $\frac{1}{8}$, $\frac{1}{4}$, board and brace measure.
6	A. Brace,	2 "	31.00	$\frac{1}{8}$, $\frac{1}{4}$, board and brace measure.
7	" B,	2 "	30.00	$\frac{1}{8}$, $\frac{1}{4}$, board measure.
0	Culls of all above,		24.00	
8	Extra,	1½ "	27.00	$\frac{1}{12}$, $\frac{1}{8}$, $\frac{1}{4}$
9	Plain,	1½ "	25.50	$\frac{1}{8}$, $\frac{1}{4}$
10	Extra, 1 foot,	1½ "	22.50	$\frac{1}{12}$, $\frac{1}{8}$, $\frac{1}{4}$
11	Plain, 1 "	1½ "	21.00	$\frac{1}{8}$, $\frac{1}{4}$
12	Cast Steel, 1 foot,	1½ "	30.00	$\frac{1}{8}$, $\frac{1}{12}$, $\frac{1}{8}$, $\frac{1}{4}$, and $\frac{1}{100}$th scale.
15	Bridge Builders',	3 "	180.00	$\frac{1}{8}$, $\frac{1}{4}$, $\frac{1}{8}$, 1 inch slot in centre.
20	Machinists', 6 inch, Nickel Plated,	1 "	13.50	$\frac{1}{16}$, $\frac{1}{12}$, $\frac{1}{8}$, $\frac{1}{4}$

Eagle Squares.

NOS.	NAMES.	WIDTH.	PRICE.	DESCRIPTION.
13	A Brace,	2 in.	$27.00	$\frac{1}{8}$, $\frac{1}{4}$, board and brace measure.
14	B,	2 "	25.50	$\frac{1}{8}$, $\frac{1}{4}$, board measure.

Iron Squares.

NOS.	NAMES.	WIDTH.	PRICE.	DESCRIPTION.
1		1½ in.	$6.00	Marked on one side.
2		1½ "	10.00	Marked on both sides.
4		2 "	14.00	Marked on both sides.

Steel Squares,			packed four dozen in a case.
Iron "			" eight " " "

The Eagle Square line of squares in Hart, Bliven & Mead's 1873 Catalogue—the bridge builder's square is listed at $180 per dozen.

Eagle Square in the 1860s had adopted the principle of diversification. In addition to iron and steel carpenters' squares and steel rules, the company made a variety of special purpose and special order squares. It made graduated steel beams for scale manufacturers, ran a lumber business, and was a manufacturer of bedsteads and cottage furniture. Its woodworking shops made upright and angle styles of boring machines, and quantities of tool handles.

William P. Kellog & Co., manufacturers of curry combs and hardware specialties, placed orders for "Polly Wog," "Dutch," and "English" style curry comb handles in quantities of twenty-five and fifty thousand at a time. The Troy Malleable Iron Works, manufacturers of Coes' pattern screw wrenches and Taft-style wrenches, ordered fifteen thousand wrench handles in various styles in August of 1847, twenty-five thousand on October 23rd of the same year, eight thousand more on the following day, and in January 1875, ten thousand extra heavy Coes's handles to be shipped at a rate of two thousand per week.

The Eagle Square firm was also early in the business of "proprietary brands" and the manufacture of squares for export. Lane, Gale & Co., of Troy, New York, hardware wholesalers and manufacturers—and one of Eagle's quantity consignees for steel squares—placed orders to South Shaftsbury for square shipments to Russell and Erwin's New York warehouse, and for 140 dozen no. 1, 2, and 4 iron squares to Peck, Stow & Wilcox of Southington, Connecticut. There are small orders for squares to be marked in both English and Spanish measure, for the foreign trade.

Levi G. Kingsley of Rutland, Vermont, on 30 July 1871, special ordered forty-eight- by thirty-six-inch squares, graduated in eighths, wanted for a slate mill. Page, Harding & Co., of Berkshire, Massachusetts, plate glass importers, in November of 1872 ordered four squares with a four-foot body, two inches wide, and two-foot tongue, one-and-one half inches wide "for glass cutting tables."

One of the steel squares made in limited quantities, and now scarce, was the "bridge builders'" square—on one or two occasions otherwise identified

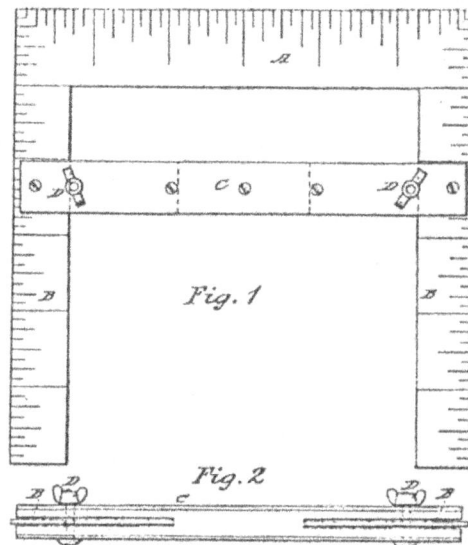

Robinson's Ruling Device

in the correspondence as a "millwright's" square.

The bridge builder's square is described in several 1870 - 1872 order letters as a square having a three-inch wide body with a twenty-inch long slot, one-inch wide, the tongue two inches in width, and "body and tongue usual length." One 1870 letter on the subject of this tool makes the following reference: "Hart Manufacturing Co.'s No. 15 is for Bridge Builder's use." The Hart, Bliven & Mead Manufacturing Co. *Catalogue and Price List* of 1873 (which displays principally the products of other manufacturers) list a line of "Steel and Iron Squares" on page 3. Though not stated, the entire list is of Eagle Square manufacture—the stock numbers correspond to those of Eagle—and here the "Bridge " is clearly described as having a "1 inch slot in centre." The reprint edition of Russell and Erwin's *Illustrated Catalogue of American Hardware*, originally published in 1865, includes at page 174 a similar list of squares, but it lacks the No. 15. This suggests that the bridge builder's model was introduced between 1865 and 1873. Both the Hart and the Russell and Erwin lists start with the No. 100, Eagle Square Company's top of the line carpenter's square.

The Eagle Square letter file records confirm that the bridge builder's square was of extra width and fabricated with a long slot on the body. A very different tool has been erroneously called a bridge builder's square in recent years. This latter square is also scarce. The patent claim states that it consists of a square having an extra tongue, to produce two permanent right angles; it is thus U-shaped, and it includes a thumb-screw-attached, end-slotted adjustable cross bar. It was patented by E. H. Robinson, Janesville, Wisconsin, 3 January 1870, and was made by Eagle Square in No. 1 and No. 2 styles with different graduations—but it is not a bridge builder's square. Unfortunately, the patent specifications for this U-shaped device describe its construction but fail to identify its function, other than to call it a "Ruling-Device." Hopefully, further investigation of the Eagle Square records will eventually bring to light more information on this tool.

D. Taft & Sons, Taftsville, Vermont

The village of Taftsville is located in the north east corner of the town of Woodstock, at the point where the town lines of Woodstock, Pomfret, Hartford, and Hartland converge. It was for almost all of the nineteenth century the site of a small Vermont industry—the making of edge tools for carpentry and agricultural purposes. One of the first settlers of the village proper was Stephen Taft, who had moved from Uxbridge, Massachusetts, at about 1793. Stephen Taft was a blacksmith and edge-tool maker. He purchased 193 acres of land and shortly thereafter constructed a dam on the Ottauquechee River as the first step in providing power for manufacturing. Daniel Taft, a younger brother who was born in Mendon, Massachusetts, on 28 December 1778 (died 30 August 1857), came to the village a year after Stephen, as an apprentice to his older brother.

Between 1793 and 1796, the brothers built a shop on the south side of the Ottauquechee and a sawmill on the north side. On 15 January 1802, Daniel started an independent business, hiring from Stephen until the following August 1st, two forge fires and two-thirds of the total time available on the trip-hammer. Two years later, Daniel, in partnership with another brother, Seth, paid Stephen $260 for the edge-tool shops and one-third of the water rights at the dam site. Stephen Taft died some time prior to the Census of 1810.

In 1811, Seth was fatally injured as a result of a fire which destroyed the shop, and Daniel Taft reconstructed the shop, purchased the balance of the water privilege, and continued the business. The census of 1820 records but one edge-tool, "Scythes," utilizing annually "2¹/₂ tones of iron, 1,500 lb grindstones, 875 pounds of steel," at a cost of $647 for raw materials, and with $40 of annual contingent expenses. One man and four individuals in the category "boys or girls" had at that time a capital investment of $647. For machinery, the Census merely lists "1 trip-hammer, etc." and notes that all machines in the plant were in operation. This data most certainly refers to the Daniel Taft edge-tool works.

Following his first marriage in 1801, Daniel Taft had three sons, Daniel, Owen, and Paschal, who joined the business—probably as they came of age. Dana's *History of Woodstock* remarks that the Taft "...scythes had the highest reputation and sold all over the country...(and) Taft's scythes always continued in the

highest repute." The business developed through the first half of the nineteenth century, doing as much as twenty thousand dollars gross per year in good times. The stock and tools of the Granger & Swan foundry of Woodstock were purchased in April 1835, and a furnace established at Taftsville for casting stoves, ploughs, boot jacks, and other products. The Taft firm was also engaged in building several Taftsville homes. Starting in 1844, when Walton's *Vermont Register and Farmer's Almanac* begins listing the manufacturing establishments in the state, "D. Taft & Sons" are identified as machinists, and as founders and manufacturers of scythes, axes, and edged tools. During this period, the firm expanded its product line, and included unusually large millwright's socket-firmer chisels which would have been sold for use in framing heavy timbers of barns and houses. A broadside illustrated in *Taftsville Tales* advertises "More-land & Nixon's New and Improved Mortising Machine, for Mortising Hubs and

Rectangular bordered mark of D. Taft, on an axe.

The imprint on a D. Taft & Sons'chisel.

Square Timber," patented in 1853, and manufactured by D. Taft & Sons.

The growth of the Taft plant can be visualized from details of the auction bill of Barnes Gilbert, Assignee of D. Taft & Sons, advertising the sale of the properties "recently owned and occupied by Daniel Taft & Sons" on 8 September 1855:

> The Shops used for the manufacture of Scythes, Axes etc, with the water power,...machinery and tools. This consisted of a Scythe and Axe Shop, of stone, 36' x 80' with an addition of 26' x 38' having a brick basement level and a framed upper story where the grinding and finishing shop was located.
>
> A wooden foundry and machine shop, 36' x 80', two stories high in front, and three stories in the rear. The lower level was the foundry and the second floor a storage area for castings. The third floor was the machine shop which housed three "Engine Lathes, one machine for planing iron, one machine for turning out boxes, one large Chucking machine, one upright Drilling machine, one machine for cutting screws, and all necessary tools."

Additional buildings in the auction included a 36 x 36 foot coal house, a 14 x 38 foot building for storing patterns, and a 38 x 50 foot, two-story frame woodworking shop with a 20-foot diameter breast wheel, three circular saws, one whip saw (probably a water-powered up-and-down saw), two planing machines, and a lathe. The water power from the dam, offered in the sale, was for two-thirds of the power—subject to a liability to provide maintenance on an equal portion of the dam itself.

The Taft factory site continued to be used by a series of operators for tool manufacture into the 1890s. Although the properties were presumably auctioned in September 1855, the published Vermont directories continued to list the D. Taft firm as functioning as late as 1858—either because the publishers had not kept up with events, or because the operation was continued briefly under the Taft name. From 1860 to 1864, C. S. Hamilton is listed as making scythes, axes, etc. at the "J. Darling Foundry" in Taftsville; yet in 1863, A. G. Dewey is recorded in the directories as the edge-tool manufacturer, with Barnes Gilbert as the proprietor of the Taftsville Foundry.

The Emerson Edge Tool Company of East Lebanon, New Hampshire (across the Connecticut River and some fifteen miles from Taftsville), operated also at the old D. Taft & Sons works from about 1874 to 1883. The company is identified in the New Hampshire directories as making "Scythes, axes, shovels,

While operating at Taftsville, the New Hampshire firm advertised in Walton's Vermont Register.

hoes, and forks (hand made)" at the East Lebanon factory into the early 1900s.

The century-long story of edge-tool production at the Taft works on the Ottauquechee apparently ends with the last use of the properties by T. M. Ryder, who is recorded as a scythe and axe manufacturer at Taftsville from 1884 to 1891.

Sources

Henry Swan Dana. *History of Woodstock, Vermont.* Boston; Houghton, Mifflin, 1899.

"Taftsville Tales," compiled and edited by Pearl G. Watson. [Taftsville] 1967.

Cass, Lewis and Aldrich and Frank R. Holmes, editors. *History of Windsor County Vermont.* Syracuse, N. Y.; D. Mason & Co., 1891.

The New Hamphire Register, and Farmers' Almanac. Concord: D. L. Guernsey. Various annual issues.

Walton's Vermont Register and Farmers' Almanac. Montpelier, E. F. Walton & Sons. Various annual issues.

A Further Note on the Bridge Builder's Square

The Eagle Square Company of South Shaftsbury, Vermont, was established on 1 January 1859. The earliest reference to a bridge builder's square that has been located in the Eagle Square Co. manuscript files at the Bailey/Howe Library of the University of Vermont was an order dated 31 July 1869. This was placed by the wholesale hardware firm of Catlin, Lane and Co., of Troy, New York, for a "Bridge Builder's Square."

New evidence of an earlier manufacturing of this unusual square is provided by an example marked "WARRANTED/STEEL/R W BANGS." Rufus W. Bangs was one of the Bennington County square makers during the 1830s to 1852, when his North Bennington square shop was destroyed by a flood. His 1833 patented rolling mill with eccentric rolls for tapering the bodies and tongues of squares (and other

tools "formerly belonging to R.W. Bangs") were part of the 1 April 1859 inventory of the then newly established Eagle Square Co.

The Bangs square was made at least seventeen years prior to the date when Eagle Square was filling the Catlin, Lane & Co. order, and eighteen years before Eagle was making a bridge builder's square as a proprietary brand for the Hart Manufacturing Company.

The example by Bangs has a twenty-four-inch long slotted body which is three inches wide, and an eighteen-inch tongue, as does the tool Eagle Square made in its own name and for Hart. One difference is that Bangs provided a brace rule on the back side of the tongue.

A. W. Whitney, Tinsmiths' Toolmaker of Woodstock, Vermont

The quiet, rather elegant setting of present-day Woodstock—like Litchfield, Connecticut, it might well be characterized today as a "parlor town"—gives but little evidence of its mid-nineteenth-century place as a manufacturing center of some consequence in Vermont industrial history.

D. Taft & Sons edge-tool manufactory flourished for many years in the Taftsville portion of the town. Woodstock supported cooper and carriage-maker shops, the dyehouse and woolen mill of Solomon Woodward, at least one foundry, a silversmith, a factory (R. Daniels & Co., later, Daniels & Raymond) that produced carding machines and other equipment for woolen mills as well as "Self-feeding" cutters of rags, rope, and straw for paper mill stock and also for cutting cattle feed, and for twenty-five years the manufacture of tinsmiths' tools and associated machinery by Aaron Warren Whitney.

Henry S. Dana, in his 1889 *History of Woodstock*, wrote that Whitney set up his shop in what had previously been a fulling mill. The Beers *Atlas of Windsor Co., Vermont* published twenty years earlier records the following under the *Woodstock Town Directory*:

> Whitney, A. W. Manfr of Whitneys Patent Tinsmiths Folder, Whitneys Patent Tinsmith Stake in Connection all Kinds of Tin & Sheet Iron Workers Machines Woodstock

Beers's detailed map of the town locates Whitney's residence on the north side of present-day Vermont Route 4 in West Woodstock, about a mile from the Woodstock town center. The machine shop was on the opposite side of the road, on the north bank of the Ottauquechee, on what was then described as the "lower flat" of the river. Whitney's "List of Prices" published in 1856, and here reproduced, offers a fine description of not only all the tinsmiths' equipment offered for sale, but also includes an inventory of what then constituted a "full set" of tinner's equipment required to practice the trade. The "Sundry Machines Not Included in the Set" were also undoubtedly made by Whitney. He made "drill stocks" (if not the ratchet drills), and one is here illustrated.

Although A. W. Whitney first appears in Walton's *Vermont Register* in 1852, where he is listed as a manufacturer of "tin machines," one may conclude that by that date he had been engaged for at least five years in making tinsmiths' tools. On 11 December 1847, he had been granted patent no. 5389 for "Iron, sheet, etc., machinery for working." A search of the Commissioner of Patents "Annual Reports" from 1837 to 1856 reveals no granting of a patent to Whitney other than no. 5389. And although the inventor identifies no less than six machines as "Patent" at the beginning of his price list, he could well have been making a legitimate claim. The improvements in design for which he received the 1847 patent were likely incorporated in all of the forming machines so described. In a similar situation, a line of turning, burring, setting down, and wiring machines shown in the 1898 Peck, Stow & Wilcox *Illustrated Catalogue and Price List of Tinsmiths' Tools and Machines* are all captioned "Raymond's Patent."

Through the period 1852 to 1870, A. W. Whitney is variously listed in Walton's registers as a machinist, or as a manufacturer of tinner's tools, tin ma-

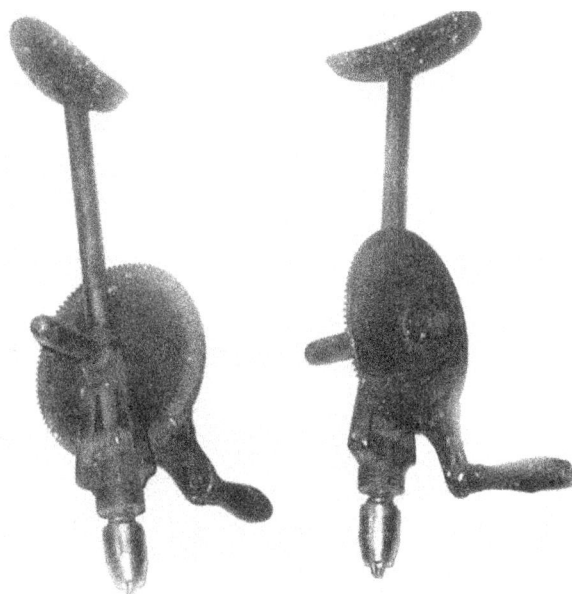

An A. W. Whitney breast drill, with modern chuck

LIST OF PRICES,

—OF—

TIN & SHEET IRON WORKERS' MACHINES,

MANUFACTURED BY

A. W. WHITNEY,

WOODSTOCK, VERMONT.

Improved 20 inch Tin Folder, or Edging Machine, warranted a better article than was ever offered to the public before. It will turn a lock from 1-16 to 3-4 inch in width with unrivalled accuracy, and with much greater despatch than any other Folder in use, **$18 00**

Improved 20 inch Groover, with Cast Steel Rolls and Iron Standard. Will swing into any position wanted, **15 00**

Patent Improved Wiring Machine, with Cast Steel Rolls, Rocking Box and Adjustable Collar, with Iron Standard that will fit any of the following Machines in the set, **11 00**

Patent Improved Large Turning Machine with Cast Steel Rolls and Rocking Box, **9 00**

Patent Improved Small Turning Machine, with Cast Steel Rolls and Rocking Box, **7 50**

Patent Improved Large Burring Machine, with Cast Steel Rolls, Rocking Box and Adjustable Collar, **7 50**

Extra Roll to Large Burring Machine, **1 00**

Patent Improved Small Burring Machine, with Cast Steel Rolls, Rocking Box and Adjustable Collar, **7 50**

Extra Roll to Small Burring Machine, **1 00**

Patent Improved Setting Down Machine with Cast Steel Rolls **7 50**

$85 00

The above comprises a full set of Tinner's Machines, which with the recent improvements, renders them far superior to any other machines in the World for durability and despatch of business. They will be furnished in full sets for cash, at a liberal discount from the above rates.

SUNDRY MACHINES NOT INCLUDED IN THE SET.

Extra Standard, same as goes with Wiring Machine. **$1 00**

No. 1 Beader for Tin, with four pair Steel Rolls and Iron Standard, *new style*, beads on 8 inches, **18 00**

Extra Rolls to No. 1 Beader, Cast Steel, per pair, **2 00**

No. 3 Beader, with four pair Steel Rolls, for tin and iron, beads on 10 inches, *new style*, **28 00**

No. 4 Beader, with four pair Steel Rolls, beads on 14 inches, for tin and iron, *new style*, **30 00**

Extra Rolls to No. 3 and 4 Beaders, Steel, per pair, **3 00**

[We have a great variety of styles of Beads, for tin, and iron, and are making constant additions to the same.]

Common Stove Pipe Former or Rollers, to fasten on bench, **16 00**

Improved Stove Pipe Former, or Rollers to fasten on bench, **28 00**

Improved Stove Pipe Former, or Rollers with iron legs, **33 00**

Common Stove Pipe Folder, **8 00**

Improved Stove Pipe Folder, similar to Tin Folder, **20 00**

Stove Pipe Groover, **20 00**

20 inch Gutter Former, with Steel Rod and Iron Tube, **4 00**

20 inch Square Pan Former, **2 50**

Square Pipe Folder, **25 00**

Guage Shears, cut 20 inches, have iron legs, operate with the foot, a new and improved article, and are warranted superior to any Guage Shear ever before offered to the public, **40 00**

Folding Machine for Can Tops, **10 00**

Green Geared Drill Stock, with four Drills, **2 50**

Bright Geared Drill Stock, with four Drills, **4 50**

Common Ratchet Drill Stock, **5 50**

No. 1 Ratchet Drill Stock, **5 00**

No. 2 Ratchet Drill Stock, **6 00**

No. 3 Ratchet Drill Stock, **7 00**

Blacking Box or Closing Machine, **16 00**

I flatter myself that I can offer such inducements as will render it an object for all that are in want of machines to purchase of me. The Rolls are all made of Cast Steel, and warranted in every respect. The improvements are such as to render it evident to any one that they must add much to their durability, and at the same time greatly facilitate the manufacture of tin and sheet iron ware. I should not use steel for rolls, did I not know it to be far more durable than iron can, by any known process, be made for such purposes. I feel sure that any person who manufactures Tin Ware cannot well do without my new Tin Folder, or one on the same principle, as it will save one-fourth the labor, and do a great deal of work that cannot be done with the old style, and will always turn both ends of the lock of the same width. The work is always held firmly against the guage during the whole operation of turning the lock, which renders it impossible for the lock to vary in width. It is made in the most thorough manner and warranted in every respect. The Guage Shears which I am about introducing to Tinsmiths, will, I trust, meet with general approbation, as those which have heretofore been manufactured have lacked durability, and rapidity in operation, which I claim to obviate. I judge from the liberal order I have recently received from the English Government, and am constantly receiving from all sections of our own country, it is not necessary for me to give any special description of the other Machines which I manufacture. I furnish Hand Tools to those who want them, of the best make in the country, and on reasonable terms. I shall be happy to receive orders from any who are in want of Machines, and will hold myself bound to do all I advertise. All persons ordering Machines will please write their name, town, county and state distinctly, so as not to have any mistake in directing goods.

A. W. WHITNEY.

Woodstock, Vt., 1856.

A. W. Whitney's 1856 "List of Prices."

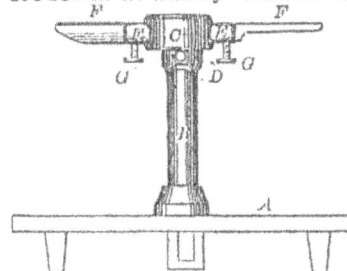

chines, tinman's machines, tinners' machines, or of machinery. In August of 1868, Whitney was granted patent no. 81,048 for a novel tinsmiths' stake which provided a head with sockets to allow interchangeable stakes such as blow horn, candle mold, creasing, etc., to be fitted to the head. Aaron was not the only member of the Whitney family involved in the development of tinsmiths' equipment. Patent no. 83,347 was issued in October of the same year to A. W. Whitney and Pardon A. Whitney for an improved sheet-metal bending machine.

In March 1870, Aaron W., Pardon A., and yet another Whitney, Foster A., collaborated in the invention of a "Tinners' and Sheet-Iron Workers' Roll." The roll described in patent no. 101,068 was "made of a single piece of cast-steel, cast for the purpose from a pattern."

Whitney's forthright self-confidence comes through to us very clearly in the final paragraph of his 1856 list of prices. It also gives a fascinating clue to the reputation for quality which he had achieved, for he refers in this same paragraph to the "liberal order I have recently received from the English Government." Many readers are perhaps familiar with the famous "Report of the Committee on Machinery of the United States of America," presented to the British House of Commons in 1855. The Committee came to the United States to purchase machinery so that the British Board of Ordnance might establish and outfit a new small arms factory (the Enfield Armoury) for production of a new pattern of rifle. They toured the northern states in 1854, visiting arms factories, U.S. government arsenals, and a number of other factories utilizing labor-saving machinery. It was with the Robbins & Lawrence machine shop and armory at Windsor, Vermont, that the Committee placed the major share of its orders for lathes, multiple-spindle drilling machines, milling, slotting, grooving, and

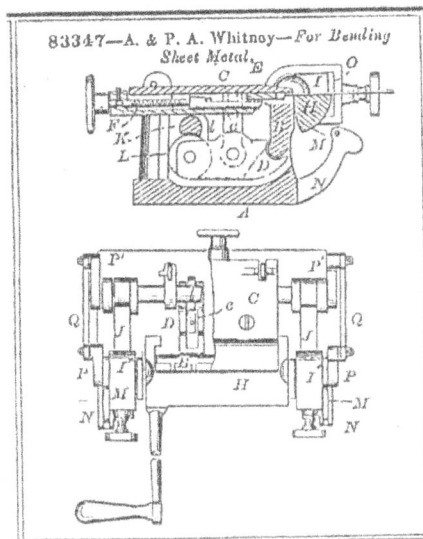

A patent (no. 83,347) was issued in 1868 to A. W. Whitney and Pardon A. Whitney for a sheet metal bending machine.

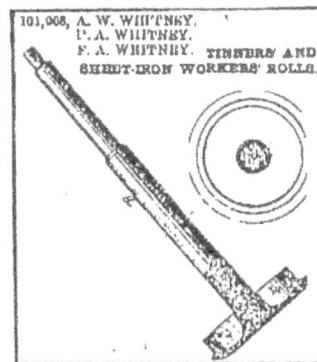

Three Whitneys, Aaron W., Pardon A., and Foster A. collaborated in this patent (no. 101,068, for a tinners' and sheet iron workers' roll.

other specialized metalworking machines (as well as twelve woodworking machines) to equip the new armory on principles embodying technology of the American system of manufactures.

While at Windsor, the Committee also learned about the manufacture of tinsmiths' machines at nearby Woodstock. Accompanied by Mr. Robbins, they visited the A. W. Whitney shop, and there placed their only order for American tinsmiths' equipment. In order to reduce the number of firms with which they would have financial transactions, the Whitney machines were bought through Robbins & Lawrence. The committee reported:

> Having visited an establishment near Windsor, where tools for tinwork are made, and obtained a list of prices, the Committee requested Messrs Robbins and Lawrence to tender to them for these machines and procure them from Mr. Whitney, the maker (as mentioned in Chapter 2, p. 115), which they did. The machines are as follows, viz.:

6	Folding machines at	$20	$120
2	Grooving machines at	16	32
4	Burring machines at	10	40
2	Wiring machines at	12	24
4	Turning machines at	10	40
2	Setting down machines at	10	20
2	No. 4 bending machines at	6	36
6	Funnel formers at	36	72
6	Hand drills at	6	36
2	Closing machines at	16	32
2	Lever shears at	50	100
			588
	Add for boxing and delivery in Boston	35	28
			$623 28

These machines to be delivered carefully packed in Boston, within six months of this date, for the sum of six hundred and twenty-three dollars, twenty-eight cents.
Signed S.W. ROBBINS, *President*
Windsor, July 26, 1854
This tender the Committee accepted the same day, viz., July 26th, 1854.

Connecticut was the major center of tinsmiths' tool manufacture through the nineteenth and into the twentieth century, with the Norths, the Peck, Smith & Co., and Peck, Stow & Wilcox familiar names. There is a connection between the Vermont and Connecticut manufacturing activity which still needs further explanation. The Peck, Smith & Co. factory in Plantsville (Southington), Connecticut, produced a machine which was stamped as C.H. Raymond's Patent, 30 August 1859. Dana's Woodstock *History* noted that Reuben Daniels, the manufacturer of machinery for woolen mills, took into partnership some years prior to the Civil War a George L. Raymond. On 8 May 1866, C.H. Raymond, who held several

patents for tinsmiths' machines from his Southington, Connecticut, address, was granted patent no. 54,599 for a tin-folding machine, at Woodstock, Vermont. George L. and C.H. Raymond may have been related, and the latter may possibly have had some association with the Whitneys. The Whitney and the Daniels shops were within a few hundred feet of each other on the Ottauquechee River.

On Christmas Eve 1864, the A.W. Whitney factory burned. Rebuilt in the summer of 1865, it was left without power by a major flood, which, according to Dana's account, washed out the dam at the factory site on 4 October 1869. Aaron W. Whitney subsequently moved to Smithville, New Jersey, and later to Stonington, Connecticut, where he died some years prior to 1889. Whitney's manufacturing plant, the nearby shop of Daniels & Raymond, and a saw mill on the bend of the Ottauquechee have disappeared with time. The residence of Aaron W. Whitney, however, still stands in West Woodstock at the location shown in the Beers "Atlas" of 1869.

Sources

"The American System of Manufactures," edited, with an Introduction by Nathan Rosenberg. Edinburgh: Edinburgh University Press, 1969.

Beers, Frederick W. *Atlas of Windsor Co., Vermont.* New York: F.W. Beers, 1869.

Dana, Henry S. *History of Woodstock, Vermont.* Boston & New York, Houghton, Mifflin, 1889.

Peck, Stow & Wilcox Co. *Illustrated Catalogue and Price List.* 1898.

U.S. Patent Office. *Annual Report of the Commissioner of Patents.* Various annual years.

Walton's Vermont Register and Farmers' Almanack. (Various years)

For further reading on tinsmiths and their products:

Demer, John H. *Jedediah North's Tinner's Tool Business.* S. Burlington, Vermont: Early American Industries Association, 1978.

DeVoe, Shirley S. *The Tinsmiths of Connecticut.* Connecticut Historical Society, 1968.

Lemuel Hedge, Vermont Inventor

In ACTIVE SCRAPBOOK no. 35, January 1981, in an article on Stearns Rules, we referred to the special machinery which was the contribution of the machinist and inventor Lemuel Hedge to the success of the E. A. Stearns Rule Company of Brattleboro. Philip Stanley of Westboro, Massachusetts, has provided an additional source of information on Hedge, an article from the 30 August 1923 issue of American Machinist by Guy Hubbard, which has been used in the following account.

Lemuel Hedge, the son of Solomon Hedge, a village blacksmith of Windsor, Vermont, was born at Windsor on 2 November 1786. After early mechanical and blacksmith training by his father, Lemuel served as an apprentice to a Windsor cabinetmaker. A few years later he established his own cabinetmaking shop in the village, and, subsequently, went into business with a partner as Hedge & Ayers. His mechanical skills and inventiveness soon turned him to other activities.

Ruled paper, in the first quarter of the nineteenth century, was laboriously produced for blank books by hand-ruling

The 1817 "Revolving Ruler," from an old woodcut

with pen, ink, and a rule. Hedge initially obtained a patent on a spring-pen ruler in 1815, and on 3 March 1817 was granted the patent for a ruling machine identified in the Secretary of State's patent list as a "Revolving ruler." These patents were issued in the name of Hedge, who was then associated with Thomas Pomroy, a local Windsor printer and stationer in whose shop the Hedge ruling pen and ruling machine were installed. Hedge's application of mechanization to the manufacture of ruled paper was a substantial success. In 1818, the *Albany Register* contained a glowing account of the operation of one of the machines at the shop of Van Vechten & Seymour, manufacturers of blank books and binders: "By means of this labour saving machine, paper can be ruled about 75 percent cheaper than heretofore. Those acquainted with the process of ruling blank books, can appreciate its value, when we inform them that two lads in the shop have repeatedly faint-lined a ream of paper on both sides in twelve minutes."

Upon the success of the machine, Pomroy and Hedge decided to abandon their respective stationery and cabinetmaking businesses, and manufacture the ruling machine exclusively. This intention was thwarted by a fire which burned to the ground the Tontine building on 25 November 1818. As a result, Hedge took to the road, selling rights on his machine in the Midwest and South. He then returned to Windsor and reverted to his cabinetmaking craft, specializing from that into making musical instruments. He secured a patent for an improvement in the bass viol and manufactured them. With the assistance of an organ builder from England, he did organ building and made wooden organ pipes. One finds an earlier instance of Hedge's versatility from *The Washingtonian* of 27 February 1815. The paper on that date (and in the aftermath of the War of 1812) carried the following advertisement: "Cork Legs —The gentlemen of the army and others who have had the misfortune to lose a leg, are informed that the subscriber makes spring cork legs on the shortest notice." Lemuel Hedge being the subscriber.

Vermont seems to have bred inventors of "dividing engines," and the many patents that were to be spawned in the South Shaftsbury area for graduating and figuring steel squares, the developments of Jeremiah Essex, Norman Millington, Dennis George, and Heman Whipple (including Millington's rule graduator of 11 February 1869), were foreshadowed by the invention of an automatic graduating engine ("Improvement in the engine for dividing scales, etc."— 1827 patent records) patented by Lemuel Hedge on 20 June 1827. Machines for the mechanical graduation of the degrees of a circle had been developed in England in the eighteenth century by Ramsden, Troughton, Bird, and others; early machines of this type were important in making surveying compasses, quadrants and sextants, and other mathematical instruments, and also in the watch and clock and other industries for accurate gear cutting. But into the nine-

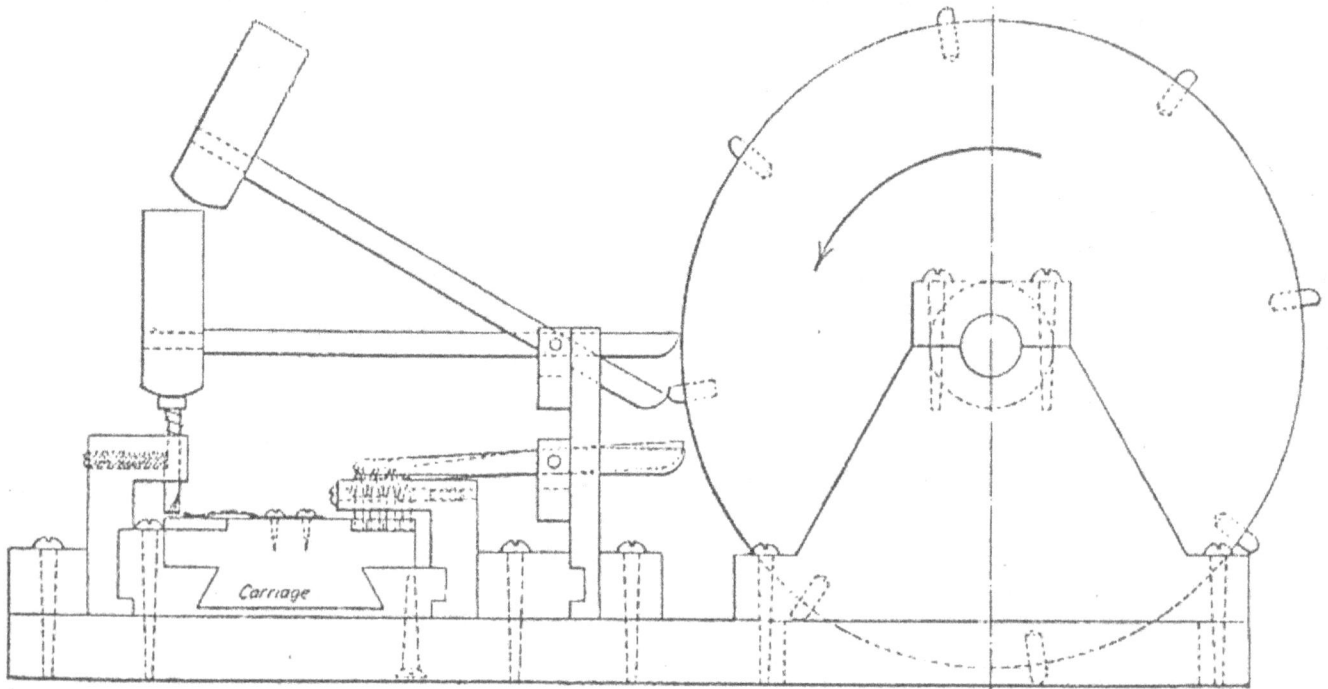

The Hedge Dividing Engine of 1827

teenth century, hand graduation of boxwood and ivory rules was the norm. The fundamental parts of Hedge's graduators were the frame, a sliding carriage, index plates, detents (i.e., stops), detent levers, marking chisels, hammers, and a cam wheel. The marking of the rules was effected by vertically sliding chisels of different widths striking the rule surface. It was possible to remove the chisels and substitute figure dies and use the machine for making the numerals, as well.

Hedge constructed his dividing engine at Windsor, and attempted to manufacture rules, but a lack of capital handicapped the venture. Published sources indicate that the manufacture of rules in Vermont was started in Brattleboro by S. Morton Clark in 1833 or

A Hedges boxwood joint rule and detail (left)

1834, and in 1835 it was moved into a factory built for Clark's firm on Birge Street. The hard times of 1837 resulted in the closing down of the business, and in 1838 Edward A. Stearns, a former Clark employee, purchased the machinery and stock, and reestablished the business. (Hubbard gives the founding of E. A. Stearns as several years earlier, "about 1830").

Hedge moved to Brattleboro to become associated with the Stearns firm some time in the 1830s and was in a substantial measure responsible for the quality and success of Stearns rules. The part of the Brattleboro factory where the graduator was in operation is said to have been off limits to all but those engaged in the mechanical scaling work. Industrial spying is by no means a twentieth-century phenomenon!

During the same period, Hedge received a patent for "Rules, Carpenters joint" on 22 April 1835. The invention of the folding rule has erroneously been attributed to Hedge, but folding rules are of ancient origin—metal folding rules having been found in archaeological excavations in Rome and in Pompeii. And it is not at all clear just what the 1835 rule joint invention was. The listing in the patent records is under the class of invention "Factory Machinery," rather than "Common Trades," which suggests that his device may not have been for the design of the rule joint, but perhaps for a machine to make the joint—either forming or hinging the brass, or sawing or routing the wood or ivory of the rule, for example.

Hubbard's article states that Hedge sold his interest in E. A. Stearns & Co. in 1840 and moved to New York City, establishing a partnership with his son, George W. Hedge, and Edwin F. Johnson, as inventor and mechanical engineer. He patented in 1842 a circular saw mill, which Hubbard reports used two saws, one above and ahead of the other. The saws were mounted on a movable carriage with the logs maintained in a stationary position.

It has also been suggested that Lemuel Hedge invented the band saw; he did not do so, but he was granted a patent, no. 6432, in May 1849 for an "Improvement in Saw-mills." His patent claim was for a means "to preserve, increase and regulate the tension of the working portion of the saw (the blade)"—that is, his contribution was a blade-tensioning device. An example of a band saw embodying his improvement was made for Hedge by the Robbins & Lawrence firm at Windsor, and demonstrated there in August 1849.

Hedge died in Brooklyn in 1854, his many patents for invention attesting to his mechanical skills and innovative concepts for machinery. Writing in 1922, Hubbard said that an original dividing engine made by Hedge on the 1827 patent was then still in use at the Stanley Rule & Level plant in New Britain, Connecticut, and was used to divide their high grade ivory rules in preference to their more modern machines.

Sources:

Cabot, Mary R. *Annals of Brattleboro*, 1681-1895. Brattleboro, Vermont: E. L. Hildreth, 1921-1922.

Hemenway, Abby M. *Vermont Historical Gazetteer*, vol. 5. Brandon, Vermont: 1891.

Hubbard, Guy. "Development of Machine Tools in New England," in *American Machinist*, 59, no. 9 (1923).

Hubbard, Guy. *Industrial History (Windsor, Vermont)*. Windsor: Town School District, 1922. pp. 23-33.

Letters from the Secretary of State transmitting a list of the names of persons to whom Patents have been issued...1817, and similar later records.

A. J. Fullam's American Stencil Tool Works

This commentary was originally published in The Chronicle *to accompany the print found on the cover of this publication.*

This month's print insert reproduces a woodcut view of a Vermont factory in the 1860s. The original type-high woodcut was copied from a photograph, probably for an advertising broadside, and is signed by the artist "Howland"—who made the woodcut. Howland was likely an employee of the Wells firm of 90 Fulton Street, New York, whose name is stamped on the verso of the printing block. The original woodcut measures $12^5/_8$ inches by $16^{11}/_{16}$ inches.

Adoniram Judson (A. J.) Fullam was born on 18 October 1835, in Ludlow, Vermont. He was educated at the Springfield Wesleyan Seminary, which had been founded in 1846. At the age of twenty-one, he was working in a machine shop in Saratoga Springs, New York, where he had gone to learn the trade. There he developed a set of tools for cutting sheet brass stencil plates. Stencils were used extensively for marking the contents of packing cases of manufactured goods packing case of manufactured goods (as shown in the foreground of the print), barrels of apples and other produce, and similar containers, as well as for addressing freight shipments.

Fullam returned to Springfield and continued to develop tools and dies for cutting brass stencils. On 10 April 1860, he was granted U. S. patent no. 27,793 for an "Improvement in Punches," and in October of the same year he secured patent no. 30,216 for an "Improved Stencil Printing Machine." He soon owned the factory building, located on the banks of the Black, at the west end of Falls Bridge. The principal products of the Stencil Tool Works were sets of punches for stencil making; these were widely advertised and sold direct to customers. A set of superior workmanship sold for twenty-five dollars, while a cheaper set was priced at ten dollars. The firm was highly successful for a short period, including the years of the Civil War, and sales were twenty thousand dollars annually. The business is listed in Atwater's *Vermont Directory and Commercial Almanac* and in Walton's *Vermont Register* during the years 1861 to 1869.

By 1868 Fullam had established the United States Piano company and developed real estate holdings in Springfield. He subsequently left Vermont and moved the piano factory to New York City, where he continued to invest in real estate. In 1895 Fullam was reported as owning fifteen houses and forty lots of city property.

The Vermont Spool & Bobbin Company

The Vermont Spool & Bobbin Co. was a firm associated with the industrial history of the Burlington, Vermont, area for a period of about eighty-four years. Established in Essex Junction in the last decade of the nineteenth century, it was reorganized and incorporated in 1896, moved to a new plant in Burlington in 1905, and ceased operations in 1977.

The manufacture of bobbins, spools, and machinery for the textile industry had early-nineteenth-century beginnings in Vermont and was stimulated by a combination of factors: a ready supply of hard maple in the forests of the state; access to streams and rivers for low-cost water power; a continuing expansion of sheep raising and its increasing supply of wool, as land was cleared in a burgeoning lumber trade; the concurrent development of yarn, thread and textile mills both locally, and, principally, in the other states of New England. Dennis George of Shaftsbury—later the first President of the Eagle Square Company—advertised his products for the textile trade in the East Bennington *Vermont Gazette* of 22 December 1846. He offered "Throstle Spindles and Fliers, Cup Spindles...and Mule and Jack Spindles." In 1849, C. E. Norris of Peacham issued an advertising broadside describing the availability of "Large Lots of Bobbins" for the woolen cloth trade at "sink or swim" prices.

The Vermont Spool & Bobbin Co. started in Essex Junction some six miles from Burlington; its founding variously dated 1893 or 1895. It was incorporated in 1896 with a capitalization of twenty-five thousand dollars. The plant was moved to Burlington in October 1905, occupying a new building on South Champlain Street, south of Maple Street, and between South Champlain and Pine Streets. Curiously, the site that was chosen for the factory was at, or close to, an L-shaped building which appears on the maps accompanying the *Burlington City Directories* as early as 1888-1889 (some sixteen years earlier) as "Bobbin Mills," immediately south of a "Marble Mill." The contemporary press fully describes the new 1905 factory as "erected on the old marble mill property." It was reported as the first building in Vermont constructed with "cement blocks"; made by workers of a firm from Port Henry, New York, a shed was constructed on the building site where the textured blocks were formed by tamping in molds, using a mix of two parts stone dust "from the Phelps ledge," one part cement, and one part sand. The blocks were cured for at least one week before use in construction. The E. F. Moore Co., which advertised at the time as "General Contractor, Roads, Street Pavements, Bridges, Dams and Heavy Masonry, Concrete construction of All Kinds a Specialty," was the builder. Construction started in May 1905, and the factory was occupied in October. The two-story building, 50 feet by 151 feet, and an estimated construction cost of

BOBBINS!

QUALITIES IMPROVED & PRICES REDUCED.

The subscriber would most respectfully say to his old customers and all engaged in the manufacture of Woolen Goods, that he has come to the determination to *put down his prices, sink or swim.*

He has now on hand LARGE LOTS of BOBBINS, and is prepared to do an extensive business to order, and is determined not to be outdone by any establishment in the country or city.

His Bobbins are Manufactured by the best machinery ever used for the business and each and every process has been carefully *watched, experimented* and *improved* on, by *skillful mechanics* for the last ten years, and no labor saving improvements left undone.

With *these* advantages and that of doing his *business* in the country where timber, water power, rent and living, are low; with all these advantages he flatters himself that he can and will furnish Manufacturers with BOBBINS at a better lay than they can be obtained elsewhere, and will guarantee, that every man will be suited as to quality and price.

N. B. Orders and samples by Mail or otherwise will be punctually attended to.

C. E. NORRIS.

Peacham, Vt. March 28, 1849.

Norris Bobbin Factory Broadside (greatly reduced). Courtesy of Special Collections Department, Bailey/Howe Library, University of Vermont.

twelve to fifteen thousand dollars, or a maximum cost per square foot of less than one dollar. Its fire resistant construction included the installation of "tested appliances for preventing fire...the most complete in this city," that is, a "wet" sprinkler system that was a novel feature for its day. The plant included a 1,560 square foot dry kiln for curing wood and a large engine and power house.

The name most continuously associated with the Spool & Bobbin Co. throughout its history was that of the Ordway family. Charles D. Ordway was President or "Manager" and owner from the early years of the century until December 1918, when his interest was sold to his son, George A. Ordway, and a group of Burlington businessmen. George A. was at various periods vice president and secretary, president and treasurer, and president. Mrs. Esther A. Ordway served as vice president and later as secretary during the 1950s and 1960s. In its final years of operation, Charles D. Ordway, grandson of the original owner, was the president.

The standard products of the company were in its later years 60 percent spools, 30 percent bobbins, and 10 percent jack spools for woolen mills. Bobbins, made chiefly for cotton and use on automatic looms, were turned in one piece from native rock maple. Spools (considerably larger—about a foot long with ends eight to ten inches in diameter) were made in three pieces, also from maple, with the two round ends of the spool screwed to the central core.

Special bobbin bits, custom machined locally, were used to drill the center holes of the bobbins and spools. Some were used in a step-boring process in which automatic drills progressively bored out the spindle holes. These drills were typically gouge-shaped spoon bits, with the normal spoon bit point or with a "parrot nose" configuration, both of which characteristically bored a straight hole in end-grained wood. The straight hole was essential to provide stable rotation of the bobbins and spools at high speeds. Again, for stable, vibration-free rotation, the circular end pieces of the large spools were turned, after assembly, by automatic lathe work which trued the circumferences and inner and outer surfaces to close tolerances. These operations eliminated the potential for any eccentric motion when in use.

Burlington and neighboring Winooski, Vermont, on the Winooski River were the sites of several successful mills from the latter part of the nineteenth century until the post World War II era. In 1907, Chace Mills No. 3, a unit of the Fall River, Massachusetts,

firm, operated eight hundred weaving machines at its Burlington mill, had thirty thousand spindles at the Winooski plant, and produced 250,000 yards of cotton goods per week. The Queen City Cotton Co., with two mills, was the largest plant in the state with over 55,000 spindles and 1,472 looms weaving twills, sateens, and cotton materials sold to printers and converters. The Winooski River Mills of the American Woolen Co. turned out worsteds, ladies' dress goods, and mens' suiting. These were typical of the local mills utilizing the products of the Spool & Bobbin Co., but the firm produced for mills throughout the country. The spools were made not only for textile thread, but also for wire and for rope, and for spooling tire cord for the Goodyear and Firestone tire companies. In the Depression year of 1933, when the plant was producing a new patented spool designed by William Wilson, the company treasurer, for Axminster carpet looms, it was reported as working at full capacity. Its period of greatest employment was during World War II, when two shifts regularly worked to fill sales requirements. From five hundred thousand to seven hundred thousand spools and bobbins per year was a normal range of output.

The decline in business and ultimate closing of the Vermont Spool & Bobbin Co. in 1977 was not the result of mills in the New England states being relocated to southern states, but rather the very substantial decline in the total number of American mills which took place in the late 1940s and the 1950s. It was a major shift in textile production that occurred on an international scale, with rapid expansion in both Europe and the Orient, which eliminated the company's markets and made the business no longer profitable.

Notes

The assistance given by Mr. Charles D. Ordway for the preparation of the Scrapbook *is gratefully acknowledged. David Blow, Special Collections Department Bailey/Howe Library, University of Vermont, kindly provided copies of pertinent newspaper articles.*

Additional Sources:

Amrhein, Joseph. *Burlington, Vermont; the Economic History of a Northern New England City.* Thesis (PhD), New York University. 1958.

Burlington City Directory, 1888-1889, 1896, etc.

Burlington Daily News Industrial Edition. 1907.

Burlington Free Press. 10 July 1905, 4 February 1939, 13 March 1969, 24 February 1971.

Salaman, R. A. *Dictionary of Tools.* 1975. (Entries for Auger, Parrot Nose; Bit, Bobbin).

Ter-Centenary of Lake Champlain. Burlington (Souvenir). 1909.

The Eccentric Taper Roller Device of Rufus W. Bangs and Stebbins D. Walbridge

The story of the development of the steel carpenters' square industry in South Shaftsbury and North Bennington, Vermont, is one in which the ingenuity and inventiveness of several individuals had a hand. They brought to these small rural communities of the state the world's major center for manufacturing carpenters' framing squares.

The contribution of Silas Hawes, who received the first U.S. patent for a carpenters' square on 15 December 1819, remains in part a mystery.[1] A Patent Office fire in Washington, probably that of December 1836, destroyed the original record. And although an act of 3 March 1837 provided for reconstructing the documentation of those patent records that had been burned, the one issued to Hawes was apparently never resubmitted for the preparation of a new record.

A Bennington County Industry

The initial impetus to the establishment of the steel square industry in South Shaftsbury and North Bennington may reasonably be attributed to Silas Hawes, his patent, and his modest factory along Paran Creek. Although he was in business in South Shaftsbury only a few years (circa 1817–1828), several local manufacturers maintained the production of squares there during the following twenty to thirty years and until the Eagle Square Company was founded on 1 January 1859. A great many of these squares were marked by the words HAWES PATENT as well as by the names of the individual makers.[2]

The second factor contributing to the successful and preeminent position of the Bennington County square industry (a position, incidentally, which has been maintained to this day by the Eagle Square Plant of Stanley Tools at South Shaftsbury) was the invention of and patenting by square makers of the locality of a series of graduating and figure-stamping machines. The graduating machines mechanically incised the inch and fractional inch markings on the steel, while figure-stamping machines marked the numerals for inches and other scales, such as the board measure. These graduating and figure stamping machines eliminated the hand process of producing on each square literally hundreds of marks, individually struck by hammer and dies, small chisels or graving tools. These inventions were patented by Jeremiah Essex, Dennis George, Norman Millington and Heman Whipple in the years 1849 to 1869.[3]

The Old Process

The manufacture of steel or iron carpenters' squares in the nineteenth century involved rolling a flat body and a tongue from metal bar stock, tapering these parts, lap-welding them together to form a right angle, grinding the surfaces smooth, jointing edges to true them to finish dimension, incising the measurement graduations and lines, stamping the figures on body and tongue, polishing, and testing for flatness, right angle, and size. The tapering of the body, or blade, (twenty-four inches long was standard) and the tongue (normally sixteen to eighteen inches in length) was done by the square-maker/blacksmith hand forging the taper after heating each part in a furnace and hammering out the taper on an anvil or by drawing out the taper under a water-powered trip hammer. With either of these methods, a certain amount of scale was formed and surface grinding was necessary to remove scale and minor surface irregularities created by heating and hammering. For over 150 years manufacturers continued to taper squares to reduce the weight of the tool at the extremities of the tongue and body. This was done to improve the "hang" of the tool in the workman's hand and made manipulation and use easier and less tiring.

The Bangs-Walbridge Patent

On 31 December 1833, a U.S. patent was issued to Rufus W. Bangs and Stebbins D. Walbridge of Bennington, Vermont, for an "improvement in Rollers, for rolling iron, steel &c." This invention, which had preceded the introduction of graduating machines by about fifteen years, was still another reason for the competitive edge enjoyed by the Bennington County square makers during the first half of the nineteenth century. Taken together with the inventions previously mentioned, it allowed the industry to maintain its favorable position well into the twentieth century. Today, in fact, the Eagle Square Plant of Stanley Tools in South Shaftsbury is responsible for the major share of the world production of steel carpenters' squares.

Rufus W. Bangs was a resident of North Bennington. He appears in the U.S. Census of 1830 and 1840 and although still a Bennington resident in the 1850s, he seems to have been skipped in the 1850 enumeration. Stebbins D. Walbridge lived a few miles

south of Bangs, in Hinsdillville, a small community also within the town lines of Bennington.[4] Ample evidence exists of carpenters' square manufacturing by Rufus W. Bangs. One example of his product, marked HAWES/PATENT/WARRANTED/STEEL/R.W. BANGS is in the collections of the Bennington (Vermont) Museum. An 1856 map of Bennington County[5] shows the residence of Bangs on West Street, probably after he had abandoned manufacturing, inasmuch as his square shop is reported to have been destroyed

Bangs & Walbridge Taper Roller Patent of 31 December 1833. The patentees suggest that their invention can be used to taper "carpenters squares, plane irons, carriage springs," etc.

Letters Patent Dated December 31st, 1833

A partial copy of the reconstructed specifications of the Bangs & Wlabridge Taper roller patent. The patentees suggest that their invention can be used to taper "carpenters squares, plane irons, carriage springs, etc.

The Schedule referred to in these Letters Patent and making part of the same, containing a description in the words of the said Rufus W. Bangs and Stebbins D. Walbridge themselves, of their improvement in Rollers, for rolling iron, steel &c.

To all persons to whom these presents shall come, Rufus W. Bangs and Stebbins D. Walbridge of Bennington, and State of Vermont, send greetings.

Be it known that we the said Rufus W. Bangs and Stebbins D. Walbridge, have invented, constructed, made and applied to use a new and useful improvement in the construction of rollers, for the rolling of iron, Steel and other metals, by which iron Steel and other metals, may be rolled of any given width, with the ends of any given difference in thickness and of a true taper in thickness from end to

end, which improvement is and may be applied to the rolling of the said metals for the construction of carpenters squares, plane irons, carriage springs and any other article or instrument requiring metals to be formed in the aforesaid shape, called the Taper roller, specified in the words following, viz:

The length which the article to be rolled is required to be made determines the size of the driving roller, which within the groove is to be in circumference of double the length of the article rolled when the required taper is from end to end, and of the precise length of the article when the taper is required to be made from the centre to the ends. The taper in the article is obtained by so constructing the...

in a flood in 1852.[6] The Patent Office record of the Bangs & Wallbridge invention suffered the same fate as that of Silas Hawes in the fire of 1836. Bangs, however, resubmitted his 1833 specifications, drawings and claims to the Washington office, and they were recorded anew in December 1842 and on 3 January 1843. They are fortunately preserved in the National Archives.[7]

The New Process

The significant feature of the patent was the provision, in a rolling mill, of one concentric roller geared to turn simultaneously with a second roller; the latter roller was eccentric and, depending on its surface length per revolution and the degree of eccentricity, would produce a taper of predetermined dimensions in the work, from one end to the other (see patent drawing). The rolls could also limit the width of the metal being rolled. Thus one pair of cast iron rolls would have been set up that could produce a 24-inch body for a square of, for example, the standard 2-inch width and incorporating a taper of from approximately $3/16$-inch thickness at one end to $5/64$-inch at the other. A second set of rolls would have been made to produce $1\frac{1}{2}$-inch wide tongues, similarly tapered. A number of passes through the rolls were probably required to obtain the required taper.

As the patent specifications state, "The length which the article to be rolled is required to be made determines the size of the driving roller, which within the groove is to be in circumference of double the length of the article rolled, when the required taper is from end to end..." The reason the eccentric roll must be twice as long as, for example, the 24-inch body for a square is that if the total circumference were only equal to 24 inches, the roller would create two tapers in one revolution.

A Lasting Improvement

The invention was indeed a significant improvement in the square manufacturing process; steel stock could be carried by tongs directly from the furnace to the rolling mill and tapers produced rapidly without hand or mechanical hammering. Because of the smooth surfaces of the rolls, significantly less grinding would be needed on the tools.

The initial 1859 Eagle Square Co. inventory reveals much concerning the production techniques of the firm.[8] It lists not only the completed squares, the ones in process, the steel, coal, grindstones, etc. "put in" by each of the four original square-making partners, but also gives details of their mechanical equipment. All of the participants (Dennis George, Jeremiah Essex, Heman Whipple, and the Hawks, Loomis Co.) list a "rolling works" and Whipple specifically lists his rolling furnace and "1 Set of smooth Rolls excentric lever cast."

As mentioned above, Rufus Bangs had ceased making squares by the time the Eagle Square Co. was founded, but the inventory shows that on 1 April 1859, Hawks, Loomis Co. put in "Tools formerly belonging to R.W. Bangs." These included "1 roll works," which would have been his eccentric roll mill.

Until 1983, some 166 years after Silas Hawes began making tapered steel squares, the Eagle Square plant was still using eccentric rolls for the tapering process. The various grades of squares now made, however, whether as Stanley products or proprietary brands such as Craftsman, are no longer tapered. The tapering added a manufacturing cost that purchasers have been unwilling to pay for, and convenience of the user has of necessity been sacrificed to the requirements of competitive pricing.

Notes

1. Paul B. Kebabian, "Early Vermont Square Makers and the Eagle Square Company," *The Chronicle*, 36:4 (December 1983), 65-66.
2. *Ibid.*, 67 includes a list of the square makers.
3. *Ibid.*, 71, note 11 gives a list of graduator and stamping machine patents.
4. Hinsdill, Joseph N., *Map, Survey, and History in brief of the town of Bennington, Vermont*, 1835. The map locates S.D. Walbridge residence.
5. Rice, E. & Harwood, C.E., *Map of Bennington County, Vermont*, (New York: C.B. Peckham, 1856), inset map of North Bennington.
6. *The Story of Shaftsbury*. (Shaftsbury, Vermont: Town Republican Centennial Committee, 1954), 31.
7. The Bangs & Walbridge patent specifications & drawings are in the National Archives, Washington, D.C. Record Group No. 241, vol. 17, pp. 464-65, and Classification Division 7917-x.
8. Inventory of the effects of Eagle Square Co. 1 January 1859. Unpublished manuscript.

Sources 4, 5, 6 and 8 are in the Special Collections Department, Bailey/Howe Library, University of Vermont, and their availability is gratefully acknowledged.

Francis A. Barrows' Vermont Plows

Castelton Corners, a small village west of the town of Castelton, Vermont, was the site of an agricultural implement factory which existed for some fifty-five years. By the time it ceased operation, the traction engine was supplanting the horse for drawing agricultural machinery.

Francis A. Barrows was born in Castelton in 1825, the son of a wagonmaker who was originally from Middlebury, Vermont. He started his factory in 1851 or 1852, producing plows for farmers in the area. His early operation was undoubtedly very small, because it is not until 1865 that he is listed in Walton's *Vermont Register* as a manufacturer of agricultural implements, with the then firm name of Barrows & Graves.

The Beers Rutland County atlas, published in 1869, includes an inset map of Castleton Corners. The Barrows Plow foundry and Barrows' home were located on the northwest corner of the intersection of North and Main Streets, while the residence of partner B. F. Graves adjoined Barrows' property.

By 1873, Graves apparently had left the partnership. The firm appears in the directories as "agricultural implements, F. A. Barrows" until 1885, though Smith and Rann's *History*, published in 1886, states that Barrows sold a half interest two years earlier (in 1883) to Simon R. Sargent. They report that "The only business of importance at Castleton Four Corners is the manufacture of agricultural implements, carried on by Francis A. Barrows since 1852. He makes about a thousand plows (including cultivators and shovel plows) annually." The authors also note that the Barrows plows "are well and favorably known at home, and are finding sale in all directions. His latest 'Clevis' is the popular plow of the country."

Hamilton Child's *Gazetteer* of 1881-1882 records that the Barrows "foundry and agricultural implement manufactory" was built in 1851 and that some thirty years later it employed six men in producing implements valued at three thousand dollars.

In 1893, Barrows was no longer associated with the firm, and S. R. Sargent & Son are listed as the proprietors. The last directory reference located for the firm is in the 1908-1909 Walton's *New Register*: "Iron Founders. Castleton. S. R. Sargent & Son."

Barrows received two U.S. patents on features of his plows: in 1878, for a mold board which was capable of adjustment laterally and vertically, and in 1884, for a "Clevis for Reversible Plows," which consisted of an adjustable rod apparently designed

294,431. CLEVIS FOR REVERSIBLE PLOWS. FRANCIS A. BARROWS, Castleton, Vt. Filed Dec. 28, 1883. (No model.)

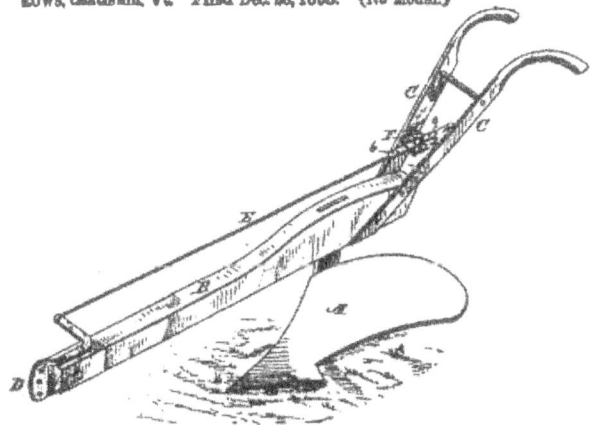

Claim.—A pivoted plow-clevis provided with a laterally-extending arm and a lever-rod connected thereto, and having at its rear end a series of holes or perforations, in combination with an angle-plate provided with a guard and a transversely-adjustable pin adapted to enter one of the series of holes or perforations in the rod, substantially as and for the purpose specified.

199,493. PLOWS. Francis A. Barrows, Castleton, Vt. Filed Dec. 28, 1877.
Brief.—The swivel-plow has a connecting-brace in two parts between the land-side and mold-board. A slot permits its extension to throw out the mold-board, and a cross-slot allows it to be raised or lowered. The mold-board is concave in front, and convex at its rear end.

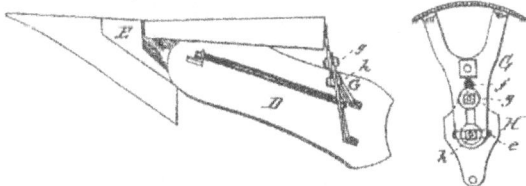

Claim.—1. A mold-board of a plow, connected to the standard or heel thereof, substantially as specified, so that the mold-board will have both a vertical and lateral adjustment, substantially as and for the purpose set forth.
2. The mold-board D, having secured thereto plate G, with vertical slot f, in combination with the plate H, having longitudinal slot e, and the bolts and nuts g h, substantially as and for the purpose specified.

22 January 1878 Patent for Plows (above), and 4 March 1884 Patent for Plow Clevis (right) issued to Francis A. Barrows. Official Gazette of the U.S. Patent Office.

to allow alteration of the vertical angle of the plow (and its operative share, colter, and mold board) as it was drawn by a horse.

Sources

Hamilton Child. *Gazetteer and Business Directory of Rutland County, Vermont,* for 1881-82.

F. W. Beers. *Atlas of Rutland County Vermont.* 1869.

Walton's Vermont Register, Farmers' Almanac, and Business Directory. Various issues.

Walton's New Vermont Register and Business Directory. 1808-1909.

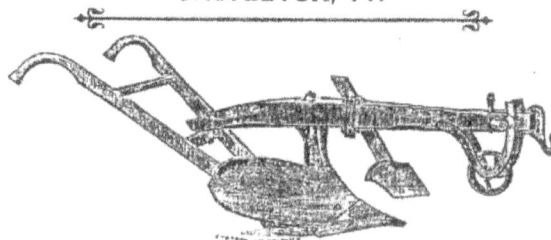

BARROWS and SARGENT
(Successors to F. A. BARROWS)
Manufacturers of Plows, Etc.
CASTLETON, VT.

Hard Metal Plow No. 1.

Easy Draft, and good on all Soils. A nice clay Plow, and the best stony land Plow. It has an adjustable beam to regulate the land. Stands 18 inches high under the beam in front of standard, and will not clog. Will be sold full-rigged with jointer or steel cutter. The jointer is valuable for inverting weeds, stubble or sod, and putting them under the soil. The above Plow is nicely polished, and is as hard as can be and he strong.

PRICES AT SHOP:

Full Rigged, with Jointer or Cutter,	$13.00
Plain, with Clevis,	10.00

ALL CUTTERS ARE SOLID STEEL ON ALL PLOWS.

REDFIELD, THE PRINTER, FAIR HAVEN, VT.

BARROWS & SARGENT,
(Successors to F. A. BARROWS)
CASTLETON, VERMONT.

MANUFACTURERS OF

THE 76 SWIVEL PLOW.
WITH STEEL AND HARD METAL MOLDBOARDS.
Patented January, 1878; also February, 1884.

REDFIELD, THE PRINTER, FAIR HAVEN, VT.

PRICES OF PLOWS:

76 A Steel Mold-board Plow, (common 2-horse)	$17 00
76 B " " (light 2-horse)	16 00
76 A Hard Metal Mold-board, polished, (common 2-horse)	15 00
76 B " " " (light 2-horse)	14 00

With Common Clevis, $1 less.

Hard Metal Plows take as nice polish as steel and will wear as long.

FIXTURES:

Steel Mold-boards,	$4 50
Hard Metal Mold-boards,	2 50
Sub Points, (that Point and Steel Mold-board bolt to)	75
76 A Points,	75
76 B Points,	75
76 C Points,	50
76 A Standard,	2 00
76 B "	1 75
76 A & B Shoes,	60
76 A & B Crotches, bolted,	65
76 A & B Finished Beams,	1 50
Solid Steel Cutters, Key and Clasp,	1 50
Finished Handles,	75

The A & B Sub Points are the same. When ordering parts of Plows, be sure and describe in every particular what it is, so that there may be no mistakes.

BARROWS & SARGENT,
Castleton, Vermont.

HAY, CORNSTALK AND STRAW
CUTTER.

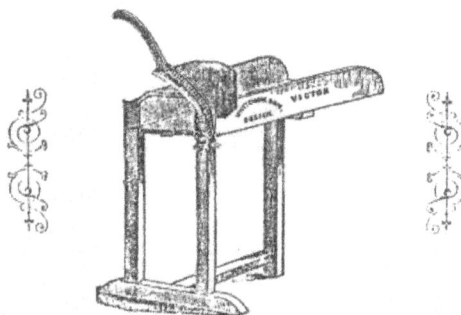

PRICE,	$7.50

No. 9 STEEL MOLD-BOARD PLOW.

Warranted to give the Best Satisfaction of any Plow in use for Clay and Sward free from Stone.

PRICE, FULL RIGGED,	$18.00.

This Plow has Points either cast or steel-laid, that cut as wide as the Plow turns, for wild grass roots, etc.

Advertising circulars of Barrows & Sargent illustrating and describing plows, a hay, cornstalk and straw cutter, and parts pists. (Illustrations substantially reduced.) Courtesy J. S. Kebabian.

The Dewey & Newton Bench Plane, and the Evidence of a Manufacturing Process

How were wooden planes made in the eighteenth and nineteenth centuries? Regrettably, very little by way of contemporary written records has come down to the present-day worker in wood, the antique tool collector, or the museum curator. One may reasonably assume that the craft of planemaking in the eighteenth century and earlier, and well into the nineteenth century, was essentially a hand process using saws, chisels, gouges, files, rasps, floats, boring implements, patterns, scribing and measuring devices, and other planes, including "mother" planes, as the task might require. In the nineteenth century, the hand processes did give way to the application of powered saws, routing machines, and other mechanical devices for a number of manufacturing operations.

Even a small country planemaking shop of circa 1851-1881, that of Jacob Lovell of Cummington, Massachusetts, had a variety of water-powered machines, including a large pitman-actuated planer.[1] In his diary for 1880, Lovell complained on 21 June that he had no water power—"cannot run the shop...half the time."

A number of descriptions of the hand planemaking processes have appeared in print in recent years, and there is a very good film, made under the direction of Ken Hawley of Sheffield, England, showing a Marples planemaker at work. These twentieth-century descriptions probably do represent the essentials of the centuries old traditional hand methods. Yet virtually nothing exists of nineteenth century sources to which one can turn to confirm assumptions about either the hand or machine processes.

There comes to mind the two-part article by Frank Wildung, then of the Shelburne Museum staff.[2] He describes the planemaking tools—chiefly chisels, gouges, and floats in the Shelburne collections—formerly used by Hartman Eckman who made bench planes, cooper's tools, and also "special" planes from boxwood, rosewood, ebony, etc., starting in 1882 at the Sandusky Tool Company. The latter special planes from exotic woods, Wildung states, were "made entirely by hand." Basing his descriptions of planemaking chiefly on information given him by William Lorenzen, a longtime employee of the Sandusky firm, Wildung makes note of a variety of machine applications for sawing and shaping bench planes, including a machine for cutting the mouth and the bed for the cutter, that was kept in a "special private room" for reasons of trade secrecy.

But aside from the conventional decorative chisel and gouge fluting on the nose and heel of planes, the hand plane chatter lines occasionally visible on the stocks of early molding planes, the circle and point marks of the center bit in the mortise sunk for inserting the bottom of the plan handle, and the marks of chisel cuts on the cutter beds and other parts of the plan throat, there is little or no evidence of a manufacturing process. One can likely make safer assumptions about the tools than the techniques.

It is of some interest, therefore, to find a wooden plane which displays clearly the marks of machining in its manufacture, much as iron planes of the nineteenth century and today show marks of foundry and milling machine work. Such a tool is the beech jointer marked with the makers' imprint H.S. DEWEY/L.W. NEWTON. (See photograph on back cover.)

The imprint on the Dewey & Newton plane.

In Atwater's *Vermont Directory and Commercial Almanac*, No. 2, 1856, there is the record of "Brown and Dewey" as manufacturers of "bench tools, piano and seraphine legs." A relatively small number of planes imprinted DEWEY & BROWN/B.FALLS.VT (i.e., Bellows Falls, in the town of Rockingham) have been found. There is no evidence in the directories that Dewey & Brown were in business before or after the year 1856. And the records of the town clerk in Bellows Falls provide no information to identify the forenames of this Bellows Falls Dewey or to connect him with Henry S. Dewey, a planemaker who appears in Bethel, Vermont (some sixty miles north of Bellows Falls) in the following year. In 1857, the Atwater *Directory* lists under Bethel the firm of Newton & Dewey as manufacturers of planes.

There is evidence from the planes themselves, however, to connect the two Dewey names. The planes made by the Dewey & Brown firm in Bellows Falls have stocks with an unusual, if not unique configuration: all four sides of the stock have a continuous, wide groove, approximately an eighth of an inch deep, that is formed below the generously rounded top edges of the plane stock. In the collection of Vermont-made woodworking tools of Mark Hughes, there are some eight bench planes of the Dewey & Brown distinctive style. Four of these are marked with the makers' imprint, while the others lack the imprint. The latter planes, moreover, all show evidence of a machine device having been used to saw out the throat cavities with small circular saws. It is almost certain that these planes were the product of Dewey at the end of his Bellows Falls manufacturing period, or at the start of his operating in Bethel, where he can be identified as Henry S. Dewey.

On 31 March 1857, U.S. Patent no. 16,954 was issued to Henry S. Dewey of Bethel, Vermont, assignor to H.S. Dewey and L.W. Newton. The patent was granted for a "Machine for Cutting the Throats of Carpenters' Plane-Stocks."

Newton was identified in the 1857 *Report of the Commissioner of Patents* as Levi W. Newton of Dorchester, Massachusetts, who moved to Bethel, possibly in 1857.

The Vermont directories carry the Newton & Dewey firm only for the year 1857. It is likely that it existed for only that one year; to date only a very few Newton & Dewey planes have been located. The two former planemakers reappear in Walton's *Vermont Register and Farmers' Almanac* in 1866 and 1867 H.S. Dewey is recorded as a Bethel millwright and machinist; in 1867, L.W. Newton is reported as a manufacturer of washing machines.

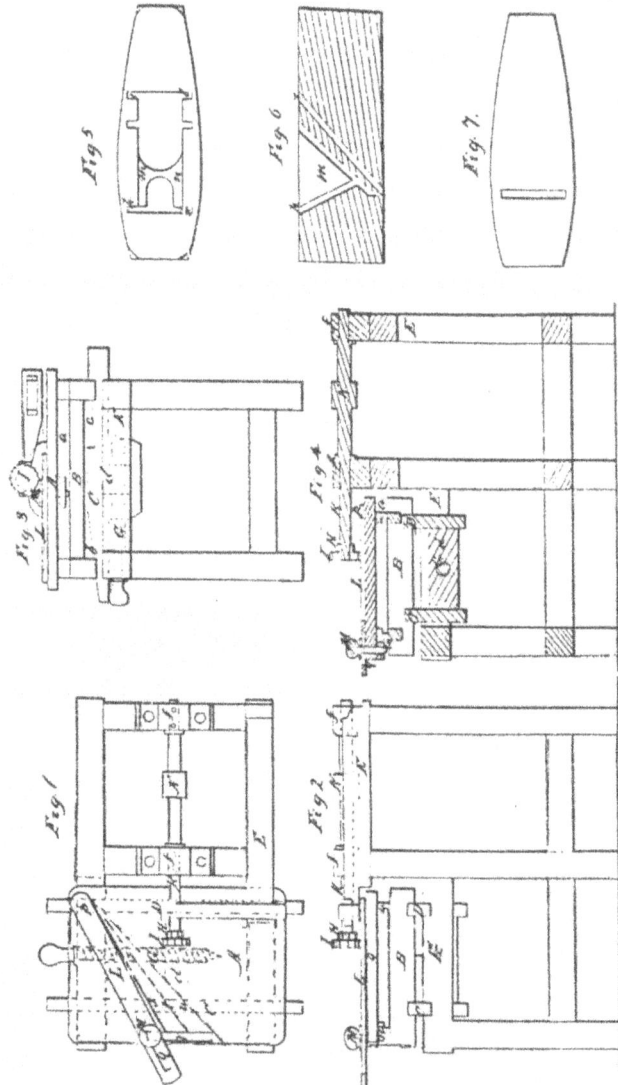

Dewey's patent drawing for the machine for cutting throats in plane stocks.

The makers' imprint on the plane suggests that the natural way to read the firm name would be "Dewey & Newton" (rather than Newton & Dewey) inasmuch as Dewey's name is at the top and that of Newton is below, upside down, and in somewhat smaller letters. This plane has a closed handle, conventional in appearance. What makes it of particular interest is the evidence that a machine, Dewey's patented "Mortising Machine," has been used in its manufacture.

Dewey describes in his patent specification the parts of his machine, and how the plane stock is placed on the movable table. The table is adjustable in height with respect to the rotary saws, to accommodate plane stocks of varying widths. A pivoted adjustable "ruler, or bearer" is used to position the plane stock at a correct angle so that it can be advanced into the saws. The saws are fixed to a rotating shaft that has a pulley to connect it to a motor source by means of an endless belt. The inventor notes that although the stocks of the smooth planes shown in figure 5 of his drawings display finished, curvilinear sides, that in "operating

The plane throat with marks of the circular saw.

with the said machine, the piece of wood of which the plane stock is to be formed should have its opposite sides parallel."

The patent states that, affixed to the shaft, a single "cutter or bur is first employed to form by successive cuts the main part of the shaving throat." In a second operation, two cutters of the shaft (as illustrated in figures 1 & 2 of the patent drawing) "effect the removal of most of the wood necessary to the formation of the mouth." The pattern of the circular saws or burrs on the cutter bed of the plane can be seen in the photograph shown here. At the center, below the slot made to accommodate the cap screw of the double iron, are the marks of the one series of circular saw cuts; a second series, wider in diameter and extending to the sides of the throat and under the cheeks, is visible below the top front of the throat.

In the light of Dewey's having received a patent on his machine, no secrecy could be attached to its construction and operation. Although the invention may have made the planemaking process somewhat more efficient by substituting the machine for hand work, by itself it was apparently not of sufficient consequence to bring success to the manufacturing firm. Of the two Dewey & Newton planes known to this writer that bear the firm imprint both show the cutting action of the invention. The machine referred to by Wildung, in use at the Sandusky Tool Company in Ohio, was certainly a device that could effect the same function of clearing wood from the stock of a wooden bench plane to form the throat. Both machines are typical examples of labor-saving mechanical devices, either ingenious or ingenuous according to one's point of view, applied to planemaking in the latter half of the nineteenth century.

The assistance of Mark Hughes in making his planes available for examination and study, is gratefully acknowledged.

Notes

1. Sayward and Streeter, "Planemaking in the Valley of the Connecticut River and the Hills of Western Massachusetts," *The Chronicle* 28, no. 2 (1975), p. 24.
2. Frank Wildung, "Making Wooden Planes in America," *The Chronicle* 8, no. 2 (1955), pp.19-21 and v. 8, no. 3 (1955), pp. 28-30.

A Brief Note on Craft Guilds and Companies and the Membership of Women

The following article was prompted by Don Johnstone of South Burlington, Vermont, submitting for the Active Scrapbook *the illustration below, with his caption "Women's Liberation?"*

The origin of the English craft guilds is in a large measure lost in the obscurity of unrecorded history. It is known that before the appearance of guilds, which incorporated skilled trades or crafts, there were religious guilds in which men and women joined together for religious and social purposes. It is not established, however, whether the later guilds, which were concerned with the training and the regulation for craft, evolved from the earlier religious organization, or whether the craft guilds grew up independently.

It is established that even in the early religious guilds, membership included both sisters and brothers, and both joined in the religious and social observances and activities. These guilds were apparently formed to provide for fellowship in life, for burial, prayers, and masses after life, and for the charitable support for any impoverished survivors. Financial support was contributed equally by both men and women. Some religious guilds were founded by women, others by men and women, with the former normally in a greater proportion.

The craft guilds were a voluntary form of association in a local, town-centered economy. They became, in time, of great significance in the economy by controlling manufacture through a period of several centuries, both on the European continent and in Britain. Some of the craft guilds went out of existence by the seventeenth century, but many developed into the later "Companies." These Companies maintained many of the social customs and traditions of the guilds, but they took on more characteristics of a closely organized capitalistic institution. They maintained a strict control of the numbers of members by approval of apprenticeship indentures, and control of standards of production by legal right to examine products (including the authority to seize and destroy inferior and "foreign" products), they set prices, and they levied quarterage—the fees or assessments paid by the member brethren and sisters. Important though the craft companies were, during the seventeenth century the advent of merchant-controlled trading organizations and the beginnings of concentration of industrial capital contributed to the decline of the Companies' power and influence in the economic structure.

The regulation of apprenticeship throughout England was established by local authority and by custom, but in 1562, during the reign of Elizabeth, the "Statute of Artificers" unified practices and conditions and brought about a national system of regulation. The statute specified a uniform term of apprenticeship by crafts (usually seven years) and that the term for artisans could not expire before age twenty-four; it established limitations on the number of apprentices (not more than three, for example, in certain specified crafts unless a journeyman were added for every additional apprentice); and it incorporated the requirement for the binding of apprentices and the enrolling or recording of the indenture document. Use of the words "Every person" and "If any person" in the language of the statute clearly indicates its applicability to both men and women as Masters or Mistresses of apprentices.

Women achieved a status as independent workers in London at an early date. The *Liber Albus* of London (1419) states that if a woman, under the authority or protection of a husband, "follows any craft within the City by herself apart, with which the husband in no way interferes, such woman shall be bound as a single woman as to all that concerns her craft."

From the beginnings of the craft guilds and companies, women were regularly admitted to membership. There is ample record of "sisters" in many of the

Companies, including the Companies of Drapers, Brewers, Clockmakers, Grocers, Stationers, Silkwomen and Throwsters of the Craft (a Company of women only), Carpenters, Joiners, Wool Merchants, and others. The 1379 poll tax returns of the West Riding of Yorkshire included among the women recorded: one farrier, one shoemaker, thirty-nine brewsters, one smith, and sixty-six websters—female weavers —of whom thirty were named Webster!

Women could become members of a Company by apprenticeship, but more often they achieved membership status by marriage to a male member. On the death of a husband, the wife could maintain her membership in the Company and was free to carry on the craft or to withdraw and transfer apprentices then indentured to another member. If she maintained the business on her own, she could receive apprentices and transfer apprentices for a consideration to another master.

The Joiners' Company of London was recognized as an independent craft in 1307. They were originally a branch within the Saddlers' Company (as were the Painters) and were responsible for constructing the wooden saddle bows and saddle-trees. Just as with union disputes of today, the Saddlers tried unsuccessfully to maintain their control over the Joiners, even to bloody street fighting. In 1571, a new Charter of the Joiners' Company of London established the geographical limits of their control to a two-mile circuit beyond the city limits. And again, just as with jurisdictional disputes of today, in 1623 the Carpenters and Joiners had to seek decisions from the London City authorities to establish schedules of the branches of woodworking exclusive to each Company.

The English system included master craftsmen, journeymen, apprentices, common laborers, husbandmen, and those of no means of subsistence who were subject to the Poor Laws. This labor hierarchy was imported to the American Colonies, including the indentured apprentice system, but the labor aristocracy fostered by the early English Guilds and Companies never became firmly established in the Colonies. The reasons why it did not can be attributed to the rapid growth and expansion of settlements, towns and cities, the mobility of individuals in the Colonies, a high demand for labor which gave skilled workers a freedom of choice in employment, and general social and economic conditions which were inhospitable to the more formal and structured patterns of life which existed in England. Although a number of Companies were established in the English tradition, such as the Carpenters' Company of the City and County of Philadelphia, effective organizations of workers in America did not occur in appreciable numbers until after the industrial pattern of the factory system had become established and trade unions emerged.

It would be satisfying if we could identify our lady woodcutter in the illustration with one of the early members of the Joiners' Company of London whom Goodman has identified in *British Plane Makers from 1700*: the widow Anna Jennion who carried on her husband's business from 1757 to 1769; Ann Wooding, widow of Robert, who maintained his planemaking shop from 1728 to 1736; Susannah Phillipson, 1761-1775; the Mistress Cain, wife of William Cain (circa 1769); the apprentice Mary Flight who was bound to William Cogdell in 1743, or her sister Sarah, apprenticed to John Jennion in the same year. Or is this one of the American planemakers noted by Roberts in *Wooden Plane Makers in Nineteenth Century America*? He has listed the widow of William Brooks (circa 1808) and Charlotte White (later Israel) working in 1840, and both of Philadelphia; Catherine Seybolt of Cincinnati, 1853-1855; and Sara S. Carpenter, widow of that superior plane maker E. W. Carpenter of Lancaster, Pennsylvania, 1860. The style of the smooth plane hanging over our subject's right shoulder in the illustration suggests a mid-nineteenth-century dating. And also one wonders just what she is doing. Is she taking her mallet and chisel to the top of the workbench?

Sources

Clark, Alice. *Working Life of Women in the Seventeenth Century.* Reprint. New York: A. Kelly, 1968.

McKee, Samuel. *Labor in Colonial New York, 1664–1776.* Port Washington, N.Y.: I. J. Friedman, 1963.

Stopes, Charlotte C. *British Freewomen.* London: Swann Sonnenschein, 1894.

Unwin, George. *The Guilds and Companies of London.* New York: Barnes & Noble, 1964.

Unwin, George. *Industrial Organization in the Sixteenth and Seventeenth Centuries.* Oxford: Clarendon Press, 1904.

Don Johnstone has suggested that we look further into the interests of our women members in the collecting and use of tools and implements,— "those associated with leather, cloth, and metal crafts as well as tools of the kitchen and sewing room." He encourages us to continue the displays and talks so successfully presented at past ACTIVE meetings, as those of Charlotte Blodgett, Edith Maher, Edith Owens, Harriet Murray, and Peg Nugent.

The Wheelwright's Tire Bolt Holder

A somewhat uncommon tool of the wheelwright, the tire bolt holder, appears as a "Whatsit" from time to time—sometimes identified but occasionally unrecognized.

After an iron or steel tire was set on the felloes of a carriage wheel or on the two-piece ash or hickory rim of a buggy or other light vehicle, the tire was bolted in place with tire bolts. These came in lengths from $1\frac{1}{2}$ to 4 inches and from $\frac{1}{8}$ to $\frac{5}{16}$ inches in diameter. The heads of these bolts were flat, and tapered to fit in countersunk holes in the iron tire. Nuts on the inner side of the felloe or rim drew the bolts tight, holding the tire in place.

A problem existed in that as a nut was tightened there was the usual tendency of the bolt to rotate and defeat tightening. Hence, the development of the Tire Bolt Holder. This wrought iron device normally had a threaded shaft that was screwed down to clamp the bolt head tight to the iron tire, preventing its rotation as the nut was tightened. Judging from the relatively few bolt holders that seem to have survived, it was most often a blacksmith-made tool. That its utility was acknowledged seems clear, however, as it also appears as a factory-made cast iron item.

The Green River firm produced a design that was tightened on the wheel by a cam lever, while a more sophisticated Reynolds device provided a crank-driven wrench, a cranked hold-down, and a cutter to shear off bolts for removing a worn tire quickly.

By 1913 an alternative solution, the Spiral Ribbed Tire Bolt had been developed to prevent the rotation of the bolt as the nut was tightened. These special tire bolts were more expensive than the standard flat head bolts. For example, in 1913 a box of 1,000 spiral ribbed bolts of the same size cost $1.30, or $0.00005 more per bolt.

The factory-made devices represent efforts to mechanize a somewhat labor intensive operation (less than two minutes claimed per bolt using the Reynolds machine), while the hand-forged bolt holders constitute examples of the ingenuity and artistry of the nineteenth century blacksmith.

Green River Tire Bolt Holder

Green River Tire Bolt Holder

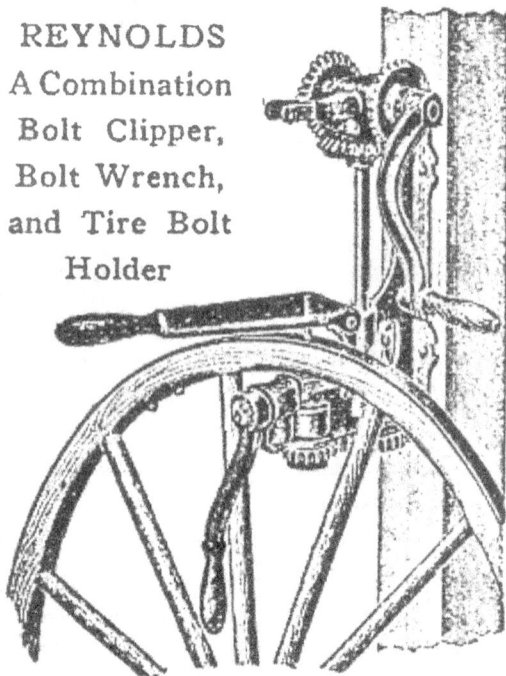

REYNOLDS A Combination Bolt Clipper, Bolt Wrench, and Tire Bolt Holder

REYNOLDS A combination Bolt Clipper, Bolt Wrench, and Tire Bolt Holder Reynolds Tire Bolting Machine

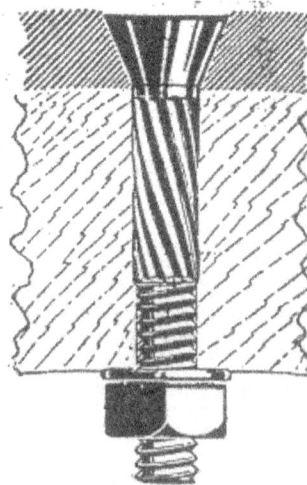

Spiral ribbed tire bolt. The catalog reads in part: "Here's That New Spiral Ribbed Tire Bolt. This is the very latest thing in tire bolt production. The spiral ribs bite into the wood and keep the bolts from turning around when putting on the nuts. Heads are nicely shaped and uniform. Threads are clean. Perfect fitting nuts. Length of rib, same on all sizes."

Sources

Cray Brothers, Cleveland, Ohio, catalogs of carriage and wagon materials and tools, 1905 and 1913 editions; catalog of the Barlow Hardware Co., Corry, Pennsylvania, 1913.

Why and Wherefore of the Axle Gauge

The axle gauge is a measuring device which is seen with relative infrequency because its need ceased with the disappearance of wagons, carriages, coaches, and other animal-drawn vehicles. The application of the tool would, however, be readily visualized by the mechanic of today familiar with the job of automotive front-end alignment.

Need for the gauge originated in a combination of design and construction characteristics: the use of a dished wheel, of a tapered axle spindle, and of a corresponding tapered or conical bore in wheel hubs. On a tapered spindle, if the wheel hub fitted thereto revolved in a perfectly horizontal plane, the rotation of the wheel on the tapered bearing would continuously exert outward pressure against the linchpin or axle nut, and wear away both the fastening and the outer end of the hub.

The first pertinent design element to note is in the construction of wheels to which "dish" was applied. "Dish" is described in Knight's *Dictionary* as, "The projection outwardly of the tire beyond the plane of the insertion of the spokes in the hub." James Arnold suggests that the dished wheel used on English farm wagons was favored because of its assumed greater strength compared with a wheel made without dish, and that the amount of dish varied from slight to deep, depending on regional tradition.

To prevent the hub from wearing against the axle nut, the tapered axle spindle was bent down so that its bottom edge would be approximately parallel to the ground. The term "swing" was used to describe this downward pitch of the spindle; it was also used to identify the resulting outward inclination of the top of the wheel. A comparable wheel attitude is provided in the automobile by adjusting the wheel for "camber." The combination of the proper degree of dish and swing would result in wheel rotation in which each spoke would be vertical as it reached the working position, and would be pitched out to a maximum degree at the top or highest point of rotation.

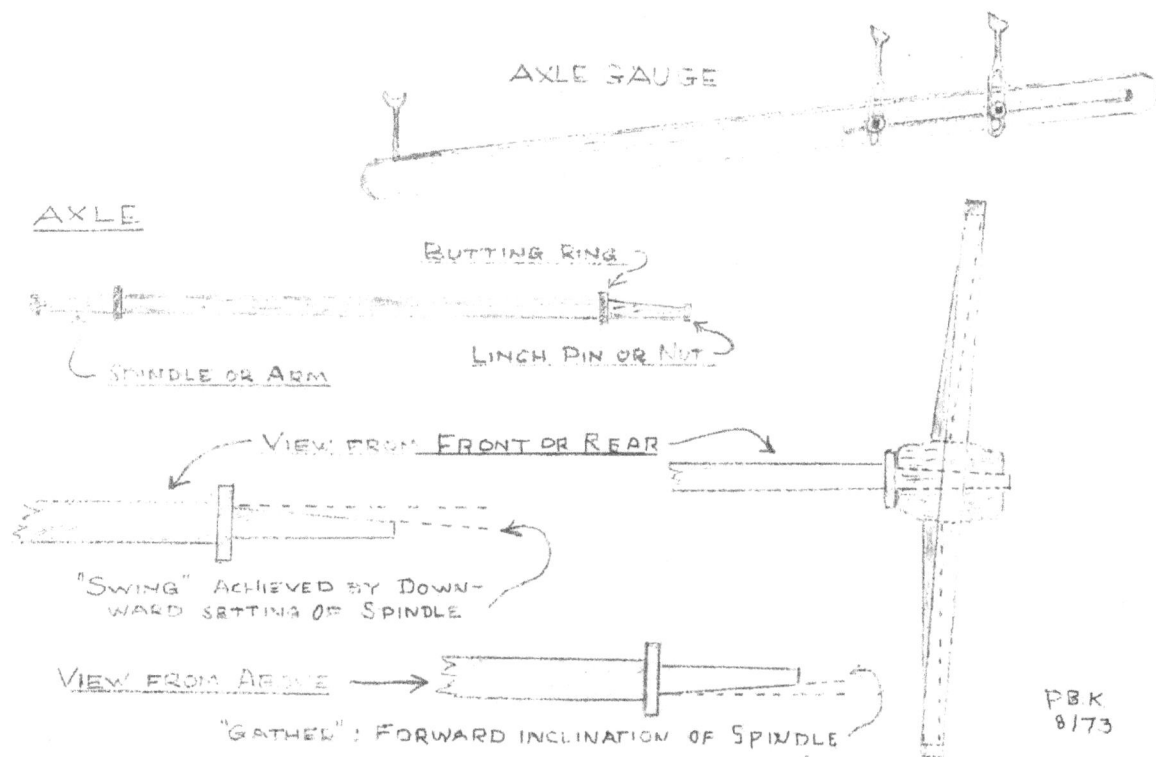

AXLE GAUGE

AXLE

BUTTING RING

SPINDLE OR ARM

LINCH PIN OR NUT

VIEW FROM FRONT OR REAR

"SWING" ACHIEVED BY DOWN-WARD SETTING OF SPINDLE

VIEW FROM ABOVE

"GATHER": FORWARD INCLINATION OF SPINDLE

PBK
8/73

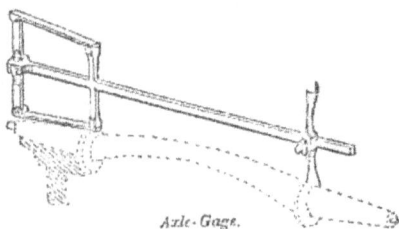

Axle-Gage.

Stratton's Axle-Gage.

Ax'le-gage. A tool by which the spindle is so adjusted in relation to the axle-tree, as to give the required *swing* and *gather*.

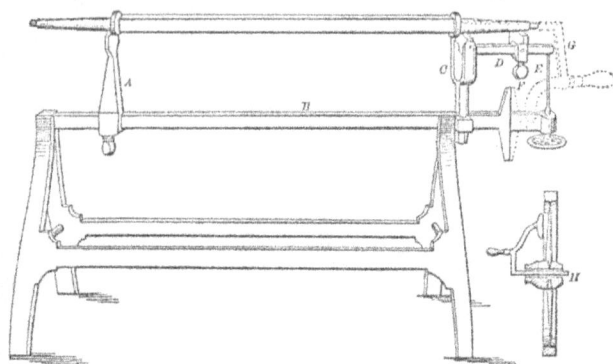

Gorton's Axle-setting Machine.

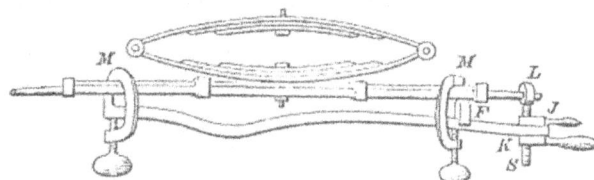

Axle-Adjuster.

Ax'le-set'ting Ma-chine'. The *Axle-setting Machine* (Fig. 506) is for setting the spindles true on the ends of the axle-trees, giving them the required *set* and *gather*.

A more portable form of the same general character is shown in the *Axle-Adjuster* (Fig. 507). It consists of a bar hooked on to the axle-tree in two places.

Further, the spindle was also inclined forward, so that its leading edge would be perpendicular to the forward direction of travel of the vehicle. "Gather" described this forward inclination, and is synonymous to the modern term "toe-in" applied to automobile wheels. Because of gather, the forward-most edge of the tire, as the wheel rotated, would tend to be centered on the hub, and the trailing edge would be pitched out and away from the vehicle body. This would also tend to put wear on the inner edge of the hub and butting ring, instead of on the axle nut, and would compensate for wear.

The axle gauge was the tool with which the blacksmith, working with the wheelwright or wainwright, measured and arrived at the proper inclination of the tapered spindles of the axle. When a non-tapered or straight cylinder spindle was used, together with a wheel made without dish, no adjustment for swing and only a very minimal adjustment for gather would have been required.

The axle gauge was placed on the axle with one arm at one butting ring, a second arm at the butting ring on the opposite end of the axle, and the third arm at the outer end of the spindle (on some gauges, a pivoted member not unlike one leg of a pair of dividers, provided the latter two reference points). The amount of downward swing to be given to a spindle would be that inclination required to result in the bottom spoke being vertical as the wheel rotated. The gauge would similarly be used to measure the amount of gather or forward inclination to be given to the spindles.

The accompanying sketches and illustrations from Knight's *Dictionary* demonstrate the positioning of the axle gauge, the axle-adjusting and setting tools used in bending the spindles to conform to requirements of the wheel being used, and swing, gather, and dish.

Sources

Knight's American Mechanical Dictionary. Boston, Houghton, Mifflin, 1882. 3 v. Entries under Axle-gage, Axle setting Machine, Dish, Swing, etc.

James, Arnold. *The Farm Wagons of England and Wales.* London, J. Baker, 1969.

The Engineer's Rule

During the nineteenth century, makers of folding boxwood rules offered these tools for sale in a wide variety of models, to serve purposes other than simple measurement of feet, inches, and fractions of inches. One rule, which could be used in performing a number of mathematical computations, was that identified as the "Engineer's" or "Engineering" rule. This was a two-foot, two-fold instrument, incorporating a logarithmic brass slide known as "Gunter's slide," together with a series of tables of "gauge points." A simpler model, called a "Carpenter's Slide Rule," provided the Gunter's slide but omitted the gauge points for performing additional calculations. The reverse side of the engineer's rule provided a twenty-four-inch measure, usually in sixteenth of an inch graduations, and additional figures and graduations for $1/4$, $1/2$ and $3/4$ -inch scales. Some makers stamped the two, over narrow edges of the rule in $1/10$- or $1/12$ -inch graduations.

The engineer's rule was apparently one of somewhat limited popularity (and no little expense) during the nineteenth century. Though quite scarce today—by comparison with other folding boxwood rules of that period—examples do turn up from time to time.

The slide incorporated in these rules has been attributed to Edmund Gunter (1581-1626), professor of astronomy in Gresham College, London. Gunter did, in fact, develop the ruled computing scale which utilized Napier's 1614 invention of logarithms, but his scale had no slide feature; computations with Gunter's lines of numbers were performed with a pair of dividers. Although invention of the *slide* rule applying Gunter's line of numbers was probably the work of the English mathematician William Oughtred, about 1630, the device has nevertheless continued to be identified as "Gunter's slide."

Of the eleven two-fold rules in the price list of E. A. Stearns & Company of Brattleboro, Vermont, probably dating from the period 1847–1849, rule no. 1 was "Arch Joint Slide, Engineering"; no. 2 was the same rule, brass-bound. How early the firm made an engineer's rule is not known, but the numbering of their engineer's rules as no. 1 and no. 2 suggests that it may have been early in their production history. The

Edward A. Stearns firm was founded by S. Morton Clark in 1833. Similarly, the 1853 *Catalogue and Invoice Prices of Rules, Planes, Gauges, Etc.* published by Hermon Chapin included examples of the twofold engineer's rule with Gunter's slide.

The first catalog of the A. Stanley & Company of New Britain, Connecticut, a predecessor to the Stanley Rule & Level Company, was published in 1855. It lists sixteen styles of the two-foot, two-fold rule. Some differed only in the structure of the joint, or hinge—round joint, square joint, or arch joint—while others incorporated functions supplementing simple linear measurement. These latter included rules with a brass slide with Gunter's scales, a table of board measure, or other kinds of functions.

When the Stanley Rule & Level Company was founded in 1857, it combined the businesses and product lines of Hall & Knapp levels and A. Stanley Company rules. The Stanley 1859 *Price List* carries the no. 6 and no. 16 engineering rules of the former A. Stanley firm. About 1863, when the E. A. Stearns Company was bought by Stanley, the Stearns rules were incorporated in the Stanley catalogs in an independent listing, retaining the Stearns name on these products. The Stanley no. 6 and no. 16 engineering rules were listed for over forty years, and the similar Stearns no. 1 and no. 2 boxwood rules were carried by Stanley at least as late as the 1979 *Price List*.

The catalog of the Chapin-Stephens Company of Pine Meadow, Connecticut, issued about 1914, lists its rule no. 16 with "Arch Joint, Bound, Engineers' Scales, Gunter Slide."

Although rules with a Gunter's slide appeared in Stanley's catalog no. 102 of 1909, its rules with the engineering data were discontinued by that date. Some time prior to 1925, rules with the Gunter's slide itself were also discontinued. One can conclude that the many improvements in slide rules in the second half of the nineteenth century—particularly the slide rule of Amédée Mannheim, the French Army officer and later professor of the École Polytechnique in Paris, made Gunter's slide obsolete.

The use of Gunter's slide and engineer's rules required an understanding of the techniques of calculation with these tools. Thus, one finds the Stanley Rule and Level Company advertising in their 1867 *Price List*, accompanying the description of slide rules, "An improved Treatise...containing full and complete instructions, enabling mechanics to make their own calculations." This was a cloth-bound book of 200 pages which sold for one dollar.

Nine years earlier, Hermon Chapin had published a small pamphlet of thirty-six pages entitled *Instructions for the Engineer's Improved Sliding Rule, with Examples of its Application*, The Chapin pamphlet explains the use of the A, B, C, and D lines, how to read and locate numbers, and standard arithmetical computation involving multiplication, division, proportions, fractions, and square and cube root. This is followed by instructions for measurement of: plane surfaces, volume of cylinders, globes, and cones, liquid measure, cask gauging, and gauging of malt in bushels, measurement and weighing of metals, etc.—based on measurements of the material or containers in feet, feet and inches, or inches. Gauge points, used in the latter computations, are listed in the columns under square, cylinder, and globe opposite the kinds of material for which the calculations are to be made, for example: wine gallons, water, gold, mercury, brass, tin, coal and marble. A table of gauge points for five- to twelve-sided polygons, and for pumping engine calculations are also stamped on to the rule.

Rules incorporating a Gunter's slide were not limited to those two-foot, two-fold rules designed for mechanics and engineers. Cask gauging, for gallon content and for computing excise tax, was a necessary function of governments. A number of slide rules were designed to serve these purposes, of which an early example is described in a book by Thomas Everard. The author invented the rule in 1863, when he was an officer in the Excise at Southampton, England. Another type of sliding rule was the "head rod," or gauging slide rule, used to measure casks, in determining gallon content. It was at one time a common tool of the gauger and excise official.

The top quality engineer's slide rule—full bound in brass with arch joint—was invariably the most expensive two-foot, two-fold boxwood rule. It was priced at thirty dollars per dozen by A. Stanley in 1855. From 1867 through 1898, the Stanley Rule and Level Company cataloged the rule at twenty-eight dollars per dozen, and, in fact, prices of rules, both in boxwood and ivory, remained virtually constant between 1867 and 1914 insofar as catalog list prices were concerned. Varying discounts undoubtedly did alter wholesale and retail prices during this period. The eighteenth- and nineteenth-century gauging tools, along with American-made folding carpenter's and mechanic's rules, have now disappeared from the rule-makers' catalogs. Few boxwood rules, other than small-caliper rules, are still manufactured today.

GAUGING.   307

CASK GAUGING.

THE operation of cask gauging is ordinarily performed with the aid of five instruments, viz., a *Gauging Slide-rule*, a *Gauging* or *Diagonal Rod*, *Callipers*, a *Bung Rod*, and a *Wantage Rod*.

THE GAUGING SLIDE-RULE.*

The Gauging Slide-rule is a flat rule, very similar to an ordinary slide-rule, except that it is not jointed, and its being adapted for use for the purpose of measuring and gauging casks; in addition to those of the ordinary computations effected by a slide-rule.

Upon the plain or outer face there are *five* lines; the first three are alike, being equally divided, and all of the same radius,† and each containing twice the length of one.

The *fourth* line is differently divided from the others, and is used in the operation of gauging, in the determination of the contents of casks when *Lying*, by the element of the depth of liquor within them, which is termed the *wet inches.*

The *fifth* line is similarly divided to the *fourth*, and is used in the operation of gauging, in the determination of the contents of casks, when *Standing*, by the element of the depth of liquor within them, which is also termed the *wet inches.*

NOTE.—The operation of gauging in this manner—that is, by the element of wet inches—is termed *Ullaging.*

Upon the opposite or inner face there are *four* lines; the *first* is divided to represent gallons,‡ the *second* is a line of mean diameters, and the *third* and *fourth* lines are divided into inches and tenths.

* As manufactured by Belcher, Brothers, & Co., New York.
† The first three lines are divided alike to the ordinary carpenters' slide-rule, or Gunter's line, described at page 297, and the operations of multiplication, division, etc., etc., may be performed, by inspection, as there described.
‡ 231 cubic inches, which is the U. S. standard gallon.

The beginning of the text on cask gauging, in which five gauging tools are identified, from Charles H. Haswell's Mensuration and Practical Geometry, *(New York, Harper & Bros., 1858).*

www.ingramcontent.com/pod-product-compliance
Lightning Source LLC
Chambersburg PA
CBHW081749200326
41597CB00024B/4451